Strip Method Design Handbook

Strip Method Design Handbook

Professor A. Hillerborg

Emeritus Professor at Lund Institute of Technology, Sweden

Routledge
Taylor & Francis Group

LONDON AND NEW YORK

First published in 1996 by E & FN Spon
First edition 1996

Published 2017 by Routledge
2 Park Square, Milton Park, Abingdon, Oxfordshire OX14 4RN
711 Third Avenue, New York, NY 10017, USA

First issued in paperback 2017

Routledge is an imprint of the Taylor & Francis Group, an informa business

A catalogue record for this book is available from the British Library

ISBN 13: 978-1-138-07561-0 (pbk)
ISBN 13: 978-0-419-18740-0 (hbk)

Contents

Notation   *xi*

Conversion factors   *xiv*

Preface   *xv*

CHAPTER 1   *Introduction*   *1*

1.1 Scope   1
1.2 The strip method   2
1.3 Strip method versus yield line theory   3
1.4 Strip method versus theory of elasticity   4
1.5 Serviceability   5
  1.5.1 Cracking   5
  1.5.2 Deformations   7
1.6 Live loads   7
1.7 Minimum reinforcement   8

CHAPTER 2   *Fundamentals of the strip method*   *9*

2.1 General   9
2.2 The rational application of the simple strip method   10
2.3 Average moments in one-way elements   10
  2.3.1 General   15
  2.3.2 Uniform loads   15
  2.3.3 Loads with a linear variation in the reinforcement direction   20
  2.3.4 Loads with a linear variation at right angles to the reinforcement direction   22
  2.3.5 Elements with a shear force along an edge   24
  2.3.6 Elements with a skew angle between span reinforcement and support   24

Contents

2.4 Design moments in one-way elements   27
    2.4.1 General considerations   27
    2.4.2 Lateral distribution of design moments   27
2.5 Design moments in corner-supported elements   29
    2.5.1 Corner-supported elements   29
    2.5.2 Rectangular elements with uniform loads   30
    2.5.3 Non-rectangular elements with uniform loads and orthogonal reinforcement   33
    2.5.4 Elements with non-orthogonal reinforcement   34
    2.5.5 Elements with non-uniform loads   34
2.6 Concentrated loads   35
    2.6.1 One-way elements   35
    2.6.2 Corner-supported elements   37
2.7 Strips   38
    2.7.1 Combining elements to form strips   38
    2.7.2 Continuous strips with uniform loads   39
2.8 Support bands   40
    2.8.1 General   40
    2.8.2 Comparison with corner-supported elements   40
    2.8.3 Application rules   42
2.9 Ratios between moments   44
    2.9.1 Ratio between support and span moments in the same direction   44
    2.9.2 Moments in different directions   45
2.10 Length and anchorage of reinforcing bars   45
    2.10.1 One-way elements   45
    2.10.2 Corner-supported elements   47
    2.10.3 Anchorage at free edges   47
2.11 Support reactions   48

CHAPTER 3   *Rectangular slabs with all sides supported*   *51*

3.1 Uniform loads   51
    3.1.1 Simply supported slabs   51
    3.1.2 Fixed and simple supports   52
3.2 Triangular loads   57
3.3 Concentrated loads   63
    3.3.1 General   63
    3.3.2 A concentrated load alone   63
    3.3.3 Distributed and concentrated loads together   68

Contents

CHAPTER 4 *Rectangular slabs with one free edge* 71

4.1 Introduction 71
    4.1.1 General principles 71
    4.1.2 Torsional moments. Corner reinforcement 73
4.2 Uniform loads 74
4.3 Triangular loads 77
4.4 Concentrated loads 82
    4.4.1 Loads close to the free edge 82
    4.4.2 Loads not close to the free edge 84

CHAPTER 5 *Rectangular slabs with two free edges* 87

5.1 Two opposite free edges 87
5.2 Two adjacent free edges 87
    5.2.1 General 87
    5.2.2 Simply supported edges, uniform loads 88
    5.2.3 One-fixed edge, uniform loads 91
    5.2.4 Two fixed edges, uniform loads 95
    5.2.5 Non-uniform loads 96

CHAPTER 6 *Triangular Slabs* 99

6.1 General 99
    6.1.1 Reinforcement directions 99
    6.1.2 Calculation of average moments in whole elements 99
    6.1.3 Distribution of reinforcement 100
6.2 Uniform loads 101
    6.2.1 All sides simply supported 101
    6.2.2 One free edge 104
    6.2.3 Fixed and simply supported edges 108
6.3 Triangular loads 109
6.4 Concentrated loads 112

CHAPTER 7 *Slabs with non-orthogonal edges* *113*

7.1 General 113
7.2 Four straight edges 114
    7.2.1 All edges supported *114*
    7.2.2 One free edge *116*
    7.2.3 Two opposite free edges *120*
    7.2.4 Two adjacent free edges *122*
7.3 Other cases 127
    7.3.1 Circular slabs with a uniform load *127*
    7.3.2 General case with all edges supported *130*
    7.3.3 General case with one straight free edge *138*
    7.3.4 General case with two or more free edges *138*

CHAPTER 8 *Regular flat slabs with uniform loads* *139*

8.1 General 139
    8.1.1 Definition of "regular" *139*
    8.1.2 Drop panels and column capitals *139*
    8.1.3 Determination of span *140*
    8.1.4 Calculation of average design moments *141*
    8.1.5 Lateral distribution of reinforcement *143*
    8.1.6 Summary of the design procedure *144*
8.2 Exterior wall or beam supports 146
    8.2.1 One single interior column *146*
    8.2.2 More than one interior column *148*
8.3 Exterior column supports 151
    8.3.1 General *151*
    8.3.2 Column support at one edge *151*
    8.3.3 Column support at a corner *154*
8.4 Slab cantilevering outside columns 155
8.5 Oblong panels and corner-supported elements 158

Contents

CHAPTER 9 *Regular flat slabs with non-uniform loads*   *161*

9.1 Introduction   161
9.2 Uniform loads in one direction   161
9.3 Different loads on panels   164
9.4 Concentrated loads   167

CHAPTER 10 *Irregular flat slabs*   *173*

10.1 General   173
10.2 Design procedure   174
10.3 Edges straight and fully supported   176
10.4 Edges straight and partly column supported   182
10.5 Edge curved and fully supported   187
10.6 Edge curved and column supported   192
10.7 Slab cantilevering outside columns   196

CHAPTER 11 *L-shaped slabs and large wall openings*   *213*

11.1 General   213
11.2 Reentrant corner   215
11.3 Supporting wall with a large opening   218
   11.3.1 Inner wall   *218*
   11.3.2 Wall along an edge   *221*
   11.3.3 Slab cantilevering outside wall   *227*

CHAPTER 12 *Openings in slabs*   *231*

12.1 General   231
12.2 Slabs with all edges supported   234
   12.2.1 Rectangular slabs   *234*
   12.2.2 Non-rectangular slabs   *242*
12.3 Slabs with one free edge   245
   12.3.1 Opening not close to the free edge   *245*
   12.3.2 Opening at the free edge   *247*

12.4 Slabs with two free edges   250
    12.4.1 Two opposite free edges   250
    12.4.2 Two adjacent free edges and simple supports   250
    12.4.3 Two adjacent free edges and fixed supports   252
12.5 Corner-supported elements   255

CHAPTER 13 *Systems of continuous slabs*   **257**

13.1 General   257
13.2 Systems of rectangular slabs   259
13.3 Rectangular slabs and concrete walls   269
13.4 Other cases   270

CHAPTER 14 *Joist floors*   **271**

14.1 General   271
14.2 Non-corner-supported floors   272
14.3 Floors with corner-supported elements   277

CHAPTER 15 *Prestressed slabs*   **285**

15.1 General   285
15.2 The simple strip method for tendons   286
15.3 Prestressed support bands   288
15.4 Flat slabs   289

References   *293*

Index   *295*

Notation

a	Width of reinforcement for distribution of a concentrated load, Section 2.6.1.
b	Width of a reinforcement band for carrying a concentrated load, Section 2.6.1.
b_a	Average width of the elements which are supported by a support band, see Section 2.8.3
c	Length in the reinforcement direction from a support to the line of zero shear force in an element, see Section 2.3. Indices are used to separate different lengths within an element, Section 2.3, or lengths belonging to different elements in the examples.
l	Width of an element, see Section 2.3. Indices are used to indicate different parts of the width.
Δl	Additional length of a reinforcing bar for anchorage beyond the point where it can theoretically be ended (Section 2.10). In Section 14.2 it has another meaning.
M	Design moment in kNm. A positive moment is a moment which causes tension in the bottom reinforcement. Indices are used for the direction of the reinforcement corresponding to the moment (x or y, sometimes also z), for support moment (s) or span moment (f), for the place where the moment is acting, e.g. a number of an element or a letter denoting a support or two letters denoting the span between two supports.
m	Design moment per unit width in kNm/m. Indices are used in the same way as for M. In general m stands for *an average moment* on the width of an element. Examples of notations are:
m_f	Span moment in the loadbearing direction.
m_{xf}	Design span moment for reinforcement in the x-direction.

Notation

m_{fl}	Design span moment in an element denoted *l*.
m_{sA}	Design support moment at support *A*.
m_{xAB}	Design span moment for reinforcement in the x-direction for the span between supports *A* and *B*.
m_{AB}	Design span moment for the span between *A* and *B*, the direction not necessarily following a coordinate axis.
Q	Load or shear force in kN/m.
q	Load per unit area in kN/m^2.
R	Reaction force in kN or kN/m.
$x, y, (z)$	Coordinates.
α	Ratio between moment in the middle strip and the average moment in a corner-supported element, Section 2.5.2.
β	Ratio between width of support strip and total width of a corner-supported element, Section 2.5.2.
γ	Factor for the determination of the length of support bars in corner-supported elements, Section 2.10.2.
———————	A free edge.
– – – – – – –	A simply supported edge.
/////////////	A fixed or continuous edge.
▱	An opening in a slab.
▰	A supporting wall.
▨	A supporting column.

A limited loaded area.

A dividing line between elements, as a rule a line of zero shear force. Design moments are, with a few exceptions, active in such lines.

The position of a support band. It may also show the position of a line of zero moment in cases where the bottom and top reinforcement have different directions.

The loadbearing direction in a one-way element. If two signs with different directions are shown within the same element it means that the load is divided between the two directions.

The two loadbearing directions in a corner-supported element.

Diagram showing the lateral distribution of a design moment. The lines within the diagram show the direction of the reinforcement and the values of the moments are written in a corresponding direction.

Diagram illustrating a load distribution.

Conversion factors

The SI-system is used throughout the book. All sizes are given in m (metres). All loads and forces are given in kN (kilonewtons), kN/m (kilonewtons per metre) or kN/m^2 (kilonewtons per square metre), depending on type of load or force. Bending moments designed M are always in kNm, bending moments designed m are always in kNm/m.

SI-units	US-units
1 m	3.281 ft
1 kN	224.8 lb.
1 kN/m	68.52 lb./ft
1 kN/m^2	20.89 psf
1 kNm	737.6 ft-lb.
1 kNm/m	224.8 ft-lb./ft

Preface

In the early fifties design methods for reinforced concrete slabs were discussed within a Swedish concrete code committee, where I was the working member, preparing the proposals. The main point of disagreement was whether the yield line theory was to be accepted in the code. Some of the committee members were against the acceptance of the yield line theory because it is in principle on the unsafe side and may lead to dangerous mistakes in the hands of designers with insufficient knowledge of its application and limitations. In the end the yield line theory was accepted with some limitations, but one of the committee members asked me if there did not exist any design method based on the theory of plasticity, but with results on the safe side. The answer that time was *No*.

Towards the end of the committee work Professor *Prager*, the well-known expert on the theory of plasticity, happened to give a series of lectures in Sweden, where I had the opportunity to get better acquainted with the two theorems of the theory of plasticity, the upper bound theorem, upon which the yield line theory is founded, and the lower bound theorem, which by then had found no practical application, at least not to reinforced concrete slabs. Both theorems were described as methods mainly intended to check the strength of a given structure, not in the first place as design methods. Also the lower bound theorem was mainly described as a basis for checking the strength of a given structure and the conclusion was that it is not very suitable for that purpose. The background to this statement was that only the application to homogenous materials like metal plates was discussed, not the application to materials where the bending strength can be varied.

It then struck me that the lower bound theorem could be used the other way round for reinforced concrete slabs, starting by seeking a statically admissible moment field and then arranging the reinforcement to take these moments. This was the beginning of the strip method. The idea was first published in a Swedish journal (in Swedish) in 1956. The theory was called *Equilibrium theory for reinforced concrete slabs*. As a special case the assumption of strips which carried the load only by bending moments was mentioned and called the *Strip method*. This is what we today call the *simple strip method*. At that time no solution existed for designing column-supported slabs by means of this equilibrium theory.

In the late fifties it was usual that slabs in Swedish apartment buildings were supported on walls plus one interior column. No suitable design method existed for this case. I was asked by the head of the design office of the Swedish firm *Riksbyggen* to propose a design method for this case. The result was a publication in Swedish in 1959, which was later translated into English by *Blakey* in Australia and published in 1964 under the title *Strip method for slabs on columns, L-shaped plates, etc.* This extension of the strip method has later become known as the *advanced strip method*.

The first time the strip method was mentioned in a non-Swedish publication was at the *IABSE* congress in Stockholm in 1960, where I presented a short paper with the title *A plastic theory for the design of reinforced concrete slabs.* This paper aroused the interest of some researchers, who studied the Swedish publications (or unofficial translations) and wrote papers and reports about the theory. Thus *Crawford* treated the strip method in his doctoral thesis at the University of Illinois, Urbana, in 1962 and in a corresponding paper in 1964.

Much early interest for the strip method was shown by *Armer* and *Wood*, who published a number of papers where the method was described and discussed. They have played a major role in making the method internationally known.

In the early seventies I had found that the interest in the method was so great that it was time to write a book which treated the method in a greater detail. The result was a book which was published in Swedish in 1974 and in English in 1975 with the title *Strip Methods of Design.* My intention with that book was twofold. I wished to show how most design problems for slabs can be treated by means of the strip method in a rigorous way, but I also wished to give advice for its practical application. Whereas I think that the first goal was reached, the second was not. The book has rightly been regarded as too theoretical and difficult for practical application.

From 1973 to my retirement I was a professor in building materials and had to devote my interest to other topics than to structural design problems. During this period I did practically nothing about the strip method except the contacts I kept with interested people.

During the last 20 years the strip method has been introduced into many textbooks on the design of reinforced concrete. In most cases the treatment is mainly limited to the basic idea and the treatment of simple cases by means of the simple strip method, as this is easiest to explain and to apply. In my opinion this is a pity, as the greatest advantage of the strip method is that it makes it possible to perform a rather simple, safe and economical design of many slabs which are complicated to design by means of other methods.

The interest in the strip method thus seems to have increased, but probably the practical application has lagged behind because of a limited understanding of the application. This made me consider the possibility of writing a new book, intended for the people in design offices. After my retirement a few years ago I have got the time for writing the book. A grant from *Åke och Greta Lissheds Stiftelse* for buying a computer and appropriate programs for that purpose has made it possible for me to carry through the project.

Whereas in my earlier book I tried to show rigorously correct theoretical solutions, this time I have allowed myself some approximations and simplifications when I have given the recommendations for the practical application. This has been done in order to simplify and systematize the numerical analyses. As far as I can judge the resulting design is always on the safe side in spite of these approximations, which sometimes cannot be shown to formally fulfil the requirements of the lower bound theorem of the theory of plasticity. Anyway the design is always safer than a design based on the yield line theory. Checks by means of the yield line theory of slabs designed according to the recommendations in this book never show that it is on the unsafe side, at least as far as I have found. A formal exception is that I have disregarded the corner levers which are sometimes taken into account in the yield line theory. Instead I have recommended lateral moment distributions where the influence of the corner levers is minimized.

The book is not intended to be read right through, but to be used in design offices as a support for the designer who meets a design problem. He should just be able to look up the type of slab and study the relevant pages in the book.

It should be pointed out that the approach in the book only gives the moments for the design of flexural reinforcement and the reaction forces, and does not give recommendations for the design with regard to shear and punching. Rules from relevant codes have to be followed in these cases.

When I started writing the book I thought that it would be a simple and straight-forward task for me to show how to apply the method. In practice it did not prove so simple when I tried to find solutions which were simple and easily explained in the more complicated cases. In spite of my efforts maybe some of the solutions still will be looked upon as complicated. It must however be remembered that many of the slabs analysed are statically complicated, e.g. flat slabs with irregularly placed columns, and that it is not realistic to hope for very simple solutions for such cases.

The book contains thousands of numerical calculations. Although I have tried to check everything thoroughly there are certainly some errors left. As all authors know it is very difficult to observe mistakes in what you have written yourself. I ask the reader to excuse possible mistakes.

I wish to express my thanks to all my friends and colleagues all around the world who by their interest and support through the years have encouraged me to decide to write this book. I refrain from mentioning names, as there is a risk that I might forget someone.

It is my sincere hope that the book will prove useful in the design offices.

Nyköping, Sweden

Arne Hillerborg

CHAPTER 1

Introduction

1.1 Scope

The general scope of this book is to give guidance on the practical application of the strip method.

The strip method is in principle a method for designing slabs so that *the safety against bending failure is sufficient.* As opposed to *the yield line theory* it gives a safe design against bending failure. The strip method does not in itself lead to a design which is close to that according to *the theory of elasticity,* nor does it take *shear* or *punching* failure into account. The additional recommendations given in this book however take the moment distributions according to the theory of elasticity into account in an approximate way and give shear forces which can be used in shear and punching design.

The strip method was first developed in the mid fifties and published in Swedish. Some translations in English were published in the sixties. These first publications showed the general principles and some applications, but they were not very complete.A more complete publication in English appeared in 1975 in the book *Strip Method of Design.* That book had a double scope: to develop theoretically well-founded rules for the application of the strip method to cases met with in practical design, and to demonstrate the application.

The development of the rules for practical application involved in many cases rather complicated discussions and theoretical derivations, which were necessary in order to prove that the resulting practical rules rested on a solid theoretical basis. As a result the book has been looked upon as theoretically complicated and difficult to read and apply. This impres-

1

sion may have been increased by its discussion of many examples of different alternative possibilities.

Bearing in mind this background and the increasing interest in the strip method, the present book has been written with the single objective of demonstrating the application to a great number of practical examples, without discussing the theoretical background in detail. Those who are interested in the theoretical background are referred to the book *Strip Method of Design*.

In order to make the application of the method to practical design as simple as possible some approximations have been used which have been estimated to be acceptable even though the acceptability has not been strictly proved. Even with these approximations the resulting designs are probably safer than many accepted designs based on yield line theory, theory of elasticity or code rules.

The intention is that a designer should be able to apply the strip method to the design of a slab met with in practice without having to read the book but just by looking for the relevant examples and following the rules given in connection with the examples, including the references to the general guidelines and rules given in the two introductory chapters.

1.2 The strip method

The strip method is based on *the lower bound theorem* of *the theory of plasticity*, which means that it in principle leads to adequate safety at the ultimate limit state, provided that the reinforced concrete slab has a sufficiently plastic behaviour. This is the case for ordinary under-reinforced slabs under predominantly static loads. The plastic properties of a slab decrease with increasing reinforcement ratio and to some extent also with increasing depth. With a design based on the recommendations in this book, including the recommendations in Section 1.5, the demand on the plastic properties of the slab is not very high. The solutions should give adequate safety in most cases, possibly with the exception of slabs of very high strength concrete with high reinforcement ratios.

As the theory of plasticity only takes into account the ultimate limit state, supplementary rules have to be given to deal with the properties under service conditions, i.e. deflections and cracks. Such supplementary rules are given in Section 1.5, and the applications of these rules are shown and sometimes discussed in the examples.

The lower bound theorem of the theory of plasticity states that if a moment distribution can be found which fulfils the equilibrium equations, and the slab is able to carry these moments, the slab has sufficient safety in the ultimate limit state. In the strip method this theorem has been reformulated in the following way:

Find a moment distribution which fulfils the equilibrium equations. Design the reinforcement for these moments.

The moment distribution has only to fulfil the equilibrium equations, but no other conditions, such as the relation between moments and curvatures. This means that many different moment distributions are possible, in principle an infinite number of distributions. Of course, some distributions are more suitable than others from different points of view. The reasons and rules for the choice of suitable distributions will be discussed in Section 1.5.

1.3 Strip method versus yield line theory

The yield line theory is based on *the upper bound theorem* of *the theory of plasticity*. This means in principle that a load is found which is high enough to make the slab fail, i.e. the safety in the ultimate limit state is equal to or lower than the intended value. If the theory is correctly applied the difference between the intended and the real safety is negligible, but there exists a great risk that unsuitable solutions may be used, leading to reduced safety factors, particularly in complicated cases like irregular slabs and slabs with free edges.

With the strip method the solution is in principle safe, i.e. the real safety factor is equal to or higher than the intended. If unsuitable solutions are used, the safety may be much higher than the intended, leading to a poor economy. From the point of view of safety the strip method has to be preferred to the yield line theory.

As the yield line theory gives safety factors equal to or below the intended value, whereas the strip method gives values equal to or above the intended value, exactly the intended value will be found in the case where the two solutions coincide. This gives the *exact solution according to the theory of plasticity*. Exact solutions should in principle be sought, as exactly the intended safety gives the best economy. How close a strip method solution is to the exact solution can be checked by applying the yield line theory to the found solution. In most of the examples in this book a check against yield line theory shows that the difference is only a few percent, which means that the strip method leads to safety factors which are equal to or just slightly above the intended values.

When comparing the strip method and the yield line theory it should be noted that the strip method is a *design method*, as a moment distribution is determined, which is used for the reinforcement design. The yield line theory is a *method for check of strength*. When the yield line theory is used for design, assumptions have to be made for the moment distribution, e.g. relations between different moments. In practice the reinforcement is often assumed to be evenly distributed, which as a matter of fact may not be very efficient. The strip method in most applications leads to a moment distribution where the reinforcement is heavier at places where it is most efficient, e.g. along a free edge or above a column support. As the strip method thus tends to use the reinforcement in a more efficient way, strip method solutions often give better reinforcement economy than the yield line solution, in spite of the fact that the strip method solution is safer. The reinforcement distribution according to the strip method solution is often also better from the point of view of the behaviour under service conditions.

A reinforcement design does not only mean the design of the sections of maximum moments, but also the determination of the lengths of reinforcing bars, and the curtailment of the reinforcement. As the strip method design is in principle based on complete moment fields, it also gives the necessary information regarding the curtailment of reinforcement. With the yield line theory it is very complicated to determine the curtailment of reinforcement in all but the simplest cases. The result from the application of the yield line theory may be either reinforcing bars which are too short or unnecessarily long bars, leading to poor reinforcement economy, as the length in practice is based on estimations, due to the complexity of making the relevant analyses.

From the above it seems evident that the strip method has many advantages over the yield line theory as a method for design of reinforced concrete slabs. In a situation where the strength of a given slab has to be checked, the yield line theory is usually to be preferred.

1.4 Strip method versus theory of elasticity

It is sometimes stated that the strip method is not a very useful practical design method today, as we are able to design slabs by means of efficient finite element programs, based on the theory of elasticity. This point of view is worth some discussion.

A finite element analysis gives a moment field, including torsional moments which also have to be taken into account for the determination of the design moments for the reinforcement. The design moment field is usually unsuitable for direct use for the design of the reinforcement. The moments have a continuous lateral variation which would require a corresponding continuous variation of the distances between reinforcing bars. This is of course not possible from a practical point of view.

One solution to this problem is to design the reinforcement for the highest design moment within a certain width. This approach is on the safe side, but may lead to poor reinforcement economy, for example, compared to a strip method solution.

A correct solution according to the theory of elasticity sometimes shows very pronounced moment concentrations. For instance, this is the case at column supports and supports at reentrant corners. It is in practice not possible to reinforce for these high local moments.

In order to avoid poor reinforcement economy and high reinforcement concentrations the reinforcement may be designed for an average design moment over a certain width. As a matter of fact this approach is based on the theory of plasticity, although applied in an arbitrary way. It may lead to results which are out of control regarding safety. In this averaging process some of the advantages of the theory of elasticity are lost. In an efficient use of finite element-based design some postprocessing procedure has to be used for the averaging. The result, e.g. regarding safety, economy, and properties in the service state, will depend on this postprocessor.

Efficient use of the finite element method, with due regard to economy and safety, may thus necessitate the use of rather sophisticated programs including postprocessors. The cost of using such programs has to be compared to the cost of making a design by means of the strip method. In most cases the time for making a strip method design by hand calculation is so short that it does not pay to use a sophisticated finite element program.

A hand calculation by means of the strip method can probably in many cases compete favourably with a design based on finite element analysis.

It should also be possible to write computer programs based on the strip method, although such programs do not so far seem to have been developed.

1.5 Serviceability

1.5.1 Cracking

In discussing cracking and crack control it is important to take into account the importance of the cracks in a realistic way. Cracks need not be avoided or limited under all conditions. Where there is no risk of reinforcement corrosion, which is the case for most indoor structures, cracks are only to be limited if they cause a visible damage. The upper surface of a slab is often covered by some flooring, carpet, parquet etc. Then a certain amount of cracking is of no practical importance and the top reinforcement may be concentrated in the parts within a section where the largest negative moments may be expected under service conditions, whereas the parts with smaller moments are left unreinforced.

In cases where cracking has to be limited there has to be sufficient reinforcement in all sections where the moments are large enough to cause cracks. This reinforcement must not yield in the service state.

The basic way of fulfilling this requirement is to choose solutions where the design moments are similar to those which may be expected according to the theory of elasticity. Some modifications of this general rule may be accepted and recommended.

It is not necessary to try to follow the elastic moments in detail regarding the lateral distribution in a section with maximum moments. The design moment may be assumed to be constant over quite large widths, even if the moments according to the theory of elasticity vary in that width. The main thing is that the average moment in the section is close to the elastic value and that there is a general agreement between the elastic and the chosen distribution.

When the moments are calculated according to the theory of elasticity it is generally assumed that the slab has a constant stiffness, independent of the amount of reinforcement. This corresponds to an assumption that there is a direct proportionality between moment and curvature. In the places where the moments are largest the curvature is also largest. From this it follows that the stresses in the reinforcement are also largest where the moments are largest, even if the reinforcement is designed for the theoretical moments. Yielding of reinforcement can be expected to ocur first in the sections where the largest moments occur if the reinforcement is designed for the moments according to the theory of elasticity. In order to avoid yielding and large cracks more reinforcement than is needed according to the theory of elasticity should in principle be chosen for the sections with the largest moments and less reinforcement should be used in sections with smaller moments. The difference between large and small design moments should thus be exaggerated compared to the values according to the theory of elasticity.

As the support moments are often larger than the span moments it may be recommended to choose a higher ratio between the numerical values of the support and span moments than according to the theory of elasticity, or at least not a smaller ratio. This recommendation also leads to a good reinforcement economy.

The main check on suitability of the design moments is thus the ratio between the numerical values of support and span moments. Where a strip with both ends fixed has a uniform load acting on its whole length between the supports this ratio should be about 2-3 from these points of view. In cases where cracks are less important on the upper surface of the slab, e.g. where there is a floor cover, values down to about 1.5 may be accepted.

Where a strip is loaded only near the ends and unloaded in the central part a higher ratio is preferred.

By the choice of a suitable ratio between support and span moments regard can also be given to the relative importance of cracks in the upper and lower surfaces of the slab for the structure in question.

Where these rules are followed the design moments according to the strip method may probably be used also for a theoretical crack control according to existing formulas.

Special attention may have to be paid to parts of a slab where the load is carried in a quite different way from that assumed in the strip method. This is particularly the case where much of the load is carried by torsional moments, but the strip method disregards the torsional moments and assumes that all the load is carried by bending moments in the directions of the coordinate axes. This situation occurs in the vicinity of corners, particularly where simply supported edges meet. It also occurs in slabs with free edges, where it may in some cases dominate.

Cracking is best limited by reinforcement which is placed approximately in the directions of the principal moments. Where large torsional moments occur these directions deviate considerably from those of the coordinate axes. Where the torsional moments dominate, the directions of the principal moments are at about 45° to the coordinate axes.

Where two simply supported edges meet at a corner the strip method in its normal application does not give any negative moment or top reinforcement. In reality there is a negative moment corresponding to a torsional moment. This moment may give cracks approximately at right angles to the bisector. The best way of limiting such cracks is by introducing some top *corner reinforcement* parallel to the bisector. The design of this reinforcement should be based on the theory of elasticity. Many codes give design recommendations.

Corner reinforcement has nothing to do with safety and it is only needed for crack control. Where cracks on the upper surface are unimportant this reinforcement may be omitted.

At corners where corner reinforcement may be needed the corner has a tendency to lift from the support. This should be taken into account either by anchoring of the corner, by arrangements that allow the corner to lift without damaging the adjacent structure, or by making an intentional crack in the upper surface.

1.5.2 Deformations

The distribution of reinforcement has a very limited influence on the deformations in the service state as long as it does not correspond to design moments which deviate appreciably from the moments according to the theory of elasticity. As long as the recommendations with regard to crack control are followed, the distribution of reinforcement may be regarded as favourable from the point of view of deformations.

Calculations of theoretical deformation values have to be based on the theory of elasticity. Most normal formulas and procedures may be applied to slabs designed by the strip method. Such analyses will not be discussed or applied in this book.

1.6 Live loads

As the strip method is based on the theory of plasticity it can only be used to give the structure the intended safety against collapse under a given constant load situation, which is normally a full load on all the structure.

Where the live load forms an important part of the total load on a continuous slab the moments at some sections may be increased by unloading some parts of the slab. Typically the increase in moments in one panel depends on the removal of loads from other panels.

The strip method can be used for analysing the change in behaviour at ultimate load due to the unloading of certain panels. This is mainly a matter of changes in requirements for lengths of reinforcing bars.

Where there is a repeated change in magnitudes and positions of live loads the stresses in reinforcement and concrete will vary. Such a variation leads theoretically to a decrease in safety against collapse through the effect known as *shake-down*. For most structures this effect is of no practical importance. In some cases it should, however, be taken into account. This has to be done by an addition to the design moments, which has to be calculated by means of the theory of elasticity.

Even though the additional moments are calculated by means of the theory of elasticity the basic moments may be calculated by means of the strip method. The additional moments are generally rather small compared to the basic moments. Approximate formulas or estimates then give an acceptable accuracy.

In cases where the change in live load magnitude is large and is repeated a great many times with nearly full intensity there may be a risk of fatigue failure. Such structures should be designed by means of the theory of elasticity.

1.7 Minimum reinforcement

Most codes contain rules regarding minimum reinforcement. These rules of course have to be followed. The rules are very different in different countries. The main reason for the great differences seems to be the lack of well-founded justification for minimum reinforcement. No account has been taken of minimum reinforcement in the examples.

Where rules for minimum reinforcement lead to more reinforcement than is needed according to an analysis, some reinforcement may be saved by making a revised analysis, where the relevant moment is increased up to a value corresponding to the minimum reinforcement. This revision will lead to a decrease in design moments at other sections with a corresponding reduction in the reinforcement requirement.

A typical example is an oblong rectangular slab. The strip method will often lead to a rather weak reinforcement in the long direction. Through a change in the positions of the lines of zero shear force this design moment can be increased while the design moment in the short direction decreases.

CHAPTER 2

Fundamentals of the strip method

2.1 General

Good introductions to the strip method are given in many textbooks, e.g. by *Ferguson*, *Breen and Jirsa*, *MacGregor*, *Nilson and Winter*, *Park and Gamble*, and *Wilby*. For a more complete presentation see *Strip Method of Design*. Here only a very short introduction will be given and the emphasis will be on rules and recommendations for practical application of the method to design.

The strip method is based on the lower bound theorem of the theory of plasticity. This means that the solutions obtained are on the safe side, provided that the theory of plasticity is applicable, which is the case for bending failures in slabs with normal types of reinforcement and concrete and normal proportions of reinforcement. As the theorem is usually formulated its purpose is to check the loadbearing capacity of a given structure. In the strip method an approach has been chosen which instead aims to design the reinforcement so as to fulfil the requirements of the theorem. The strip method is thus based on the following formulation of the lower bound theorem:

Seek a solution to the equilibrium equation. Reinforce the slab for these moments.

It should be noted that the solution has only to fulfil the equilibrium equation, but not to satisfy any compatibility criterion, e.g. according to the theory of elasticity. As a slab is highly statically indeterminate this means that an infinite number of solutions exist.

The complete equilibrium equation contains bending moments in two directions, and torsional moments with regard to these directions. Any solution which fulfils the equation can, in principle, be used for the design, and thus an infinite number of possible designs exist. For practical design it is important to find a solution which is favourable in terms of economy and of behaviour under service conditions.

From the point of view of economy, not only is the resulting amount of reinforcement important, but also the simplicity of design and construction. For satisfactory behaviour under service loading the design moments used to determine the reinforcement should not deviate too much from those given by the theory of elasticity.

Torsional moments complicate the design procedure and also often require more reinforcement. Solutions without torsional moments are therefore to be preferred where this is possible. Such solutions correspond to *the simple strip method*, which is based on the following principle:

In the simple strip method the load is assumed to be carried by strips that run in the reinforcement directions. No torsional moments act in these strips.

The simple strip method can only be applied where the strips are supported so that they can be treated like beams. This is not generally possible with slabs which are supported by columns, and special solution techniques have been developed for such cases. One such technique is called *the advanced strip method*. This method is very powerful and simple for many cases encountered in practical design, but as hitherto presented it has had the limitation that it requires a certain regularity in slab shape and loading conditions. It has here been extended to more irregular slabs and loading conditions.

An alternative technique of treating slabs with column supports or other concentrated supports is by means of the simple strip method combined with *support bands*, which act as supports for the strips, see Section 2.8. This is the most general method which can always be applied and which must be used where the conditions that control the use of other methods are not met. It requires a more time-consuming analysis than the other methods.

2.2 The rational application of the simple strip method

In the simple strip method the slab is divided into strips in the directions of the reinforcement, which carry different parts of the total load. Usually only two directions are used, corresponding to the x- and y-directions. Each strip is then considered statically as a one-way strip, which can be analysed with ordinary statics for beams.

The load on a certain area of the slab is divided between the strips. For example, one half of the load can be taken in one direction and the other half in another direction. Generally, the simplest and most economical solution is, however, found if the whole load on each area is carried by only one of the strip directions. This principle is normally assumed in this book. We can thus formulate the following principle to be applied in most cases:

The whole load within each part of the slab is assumed to be carried by strips in one reinforcement direction.

In the figures the slab is divided into parts with different loadbearing directions. The relevant direction within each area is shown by a double-headed arrow (see Fig. 2.2.1).

The load is preferably carried with a minimum of cost, which normally means with a minimum amount of reinforcement. As a first approximation this usually means that the load should be carried in the direction that runs towards the nearest support, as this results in the minimum moment and the minimum reinforcement area. From the point of view of economy, the lengths of the reinforcing bars are also important. Where the moments are positive, the length of the bars is approximately equal to the span in the relevant direction. In such cases, therefore, more of the load should be carried in the short direction in a rectangular slab.

A consequence of these considerations is that a suitable dividing line between areas with different loadbearing directions is a straight line which starts at a corner of a slab and forms an angle with the edges. Fig. 2.2.1 shows a typical simple example, a rectangular slab with a distributed load and more or less fixed edges. The dividing lines are shown as dash-dot lines.

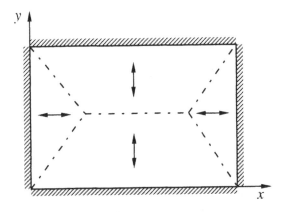

Fig. 2.2.1

The dividing lines are normally assumed to be *lines of zero shear force*. Along these lines the shear force is thus assumed to be zero (in all directions). The use of lines of zero shear force makes it possible to simplify and rationalize the design at the same time, as it usually leads to good reinforcement economy. For the choice of positions of the lines of zero shear force the following recommendations may be given.

A line of zero shear force which starts at a corner where two fixed edges meet may be drawn approximately to bisect the angle formed by these edges, but maybe a little closer to a short than to a long edge.

A line of zero shear force which starts at a corner where two freely supported edges meet should be drawn markedly closer to the shorter edge. The distances to the edges may be chosen to be approximately proportional to the lengths of the edges, for example.

Where a fixed edge and a freely supported edge meet, the line of zero shear force should be drawn much closer to the free edge than to the fixed edge.

The economy in reinforcement is not much influenced by variations in the positions of the lines of zero shear force in the vicinity of the optimum position. In cases of doubt it is fairly easy to make several analyses with different assumptions and then to compare the results.

The use of lines of zero shear force is best illustrated on a simple strip (or a beam). The slab strip in Fig. 2.2.2 is acted upon by a uniform load q and has support moments m_{s1} and m_{s2} (shown with a positive direction, though they are normally negative). The corresponding lines for shear forces and moments are also shown. The maximum moment m_f occurs at the point of zero shear force.

The parts to the left and right of the point of zero shear force can be treated as separate *elements* if we know or assume the position of this point. These separated elements are shown in the lower figure. The following equilibrium equation is valid for each of the elements:

$$m_f - m_s = \frac{qc^2}{2} \qquad (2.1)$$

with the indices 1 and 2 deleted.

The beam in Fig. 2.2.2 can thus be looked on as being formed by two elements, which meet at the point of zero shear force. This corresponds to strips in the y-direction in the central part of the slab in Fig. 2.2.1.

In many cases the loaded elements of the strips do not meet, but instead there is an unloaded part in between. This is the case for thin strips in the x-direction in Fig. 2.2.1. Such a strip is illustrated in Fig. 2.2.3. It can be separated into three elements, the loaded elements near the ends and the unloaded element in between. The unloaded element is subjected to zero shear force, and thus carries a constant bending moment m_f. The span moments m_f must be the same at the inner ends of the two load-bearing elements in order to maintain equilibrium.

In a rigorously correct application of the simple strip method we would study many thin strips in the x-direction in Fig. 2.2.1 with different loaded lengths and different resulting design moments. This leads to a very uneven lateral distribution of design moments and a reinforcement distribution which is unsuitable from a practical point of view. For practical design the average moment over a reasonable width must be considered. In the first place, therefore, the analysis should give the average moments.

To calculate these average moments we introduce *slab elements*, which are the parts of the slab bordered by *lines of zero shear force* and one supported edge. Each slab element is

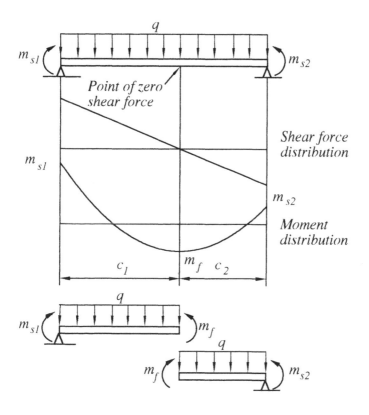

Fig. 2.2.2

actively carrying load in one reinforcement direction, the direction shown by the double-headed arrow. Thus the slab in Fig. 2.2.1 is looked upon as consisting of four slab elements, two active in the x-direction and two in the y-direction. The load within each element is assumed to be carried only by bending moments corresponding to the reinforcement direction. Such elements are called *one-way elements*. Each one-way element has to be supported over its whole width. The average moments and moment distributions are discussed in Sections 2.3-5. Cases which can be treated by means only of one-way elements can be found by the simple strip method.

Even though the dividing lines between elements with different load-bearing directions, shown as dash-dot lines, are normally lines of zero shear force, there are occasions when the analysis is simplified by using such lines where the shear force is not zero. It is then on the

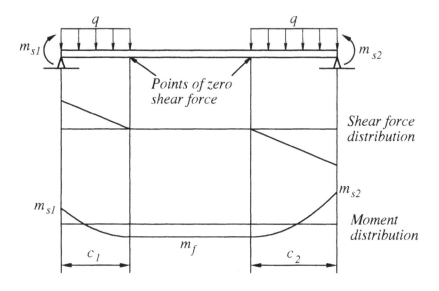

Fig. 2.2.3

safe side to undertake the analysis as if the shear force were zero along such lines, provided that the strip of which the element forms part has a support at both ends.

This is explained in Fig. 2.2.4, which shows a strip loaded only in the vicinity of the left end. The loaded part corresponds to an element. The right-hand end of the loaded part corresponds to a dividing line between elements. The upper curve shows the correct moment curve, whereas the lower curve is determined on the assumption that the shear force is zero at the dividing line. The moments according to the lower curve are always on the safe side compared to those according to the correct curve. Where this approximation is used the economical consequences for the reinforcement are usually insignificant.

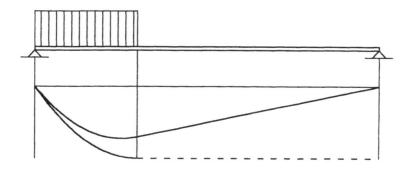

Fig. 2.2.4

2.3 Average moments in one-way elements

2.3.1 General

The load on a one-way element is carried to the support by bending moments in one direction, corresponding to the direction of the main reinforcement. This direction is shown in figures as a short line with arrows at both ends, see e.g. Fig 2.2.1. This type of arrow thus indicates that the element is a one-way element and also shows the direction of the load-bearing reinforcement.

A one-way element is supported only along one edge. Normally the shear force is zero along all the other edges. The formulas given below refer to this case. The edges with zero shear force are shown as dash-dot lines.

In most cases the reinforcement direction is at right angles to the supported edge. This case will be treated first.

The load per unit area is denoted q.

Theoretically, a one-way element consists of many parallel thin strips in the reinforcement direction. The moment in each thin strip can be calculated and a lateral moment distribution can thus be determined. The principle is illustrated in Fig. 2.3.1, which shows a triangular element with load-bearing reinforcement in the x-direction and with one side parallel to that direction.

The element carries a uniform load q and is divided into thin strips in the x-direction. Each strip is assumed to have zero shear force at the non-supported end. The length of a strip is yc/l, and the sum of end moments in each thin strip can be written

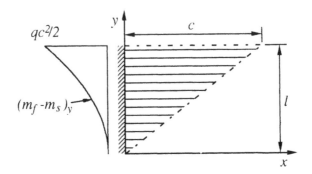

Fig. 2.3.1

$$(m_f - m_s)_y = \frac{q}{2}\left(\frac{yc}{l}\right)^2 \tag{2.2}$$

The corresponding lateral moment distribution is shown in the figure. The *average moment* is $qc^2/6$.

Formulas are given below for average moments in typical elements with distributed loads. These formulas are utilised in the numerical examples. The theoretical moment distributions are illustrated. In practical design applications the moment distribution is simplified as discussed in Section 2.4.

2.3.2 Uniform loads

In a *rectangular one-way element*, Fig. 2.3.2, the sum of average moments is

$$m_f - m_s = \frac{qc^2}{2} \tag{2.3}$$

In a *triangular one-way element*, Fig. 2.3.3, the sum of average moments is

$$m_f - m_s = \frac{qc^2}{6} \tag{2.4}$$

In a *trapezoidal one-way element*, Fig. 2.3.4, the sum of average moments is

$$m_f - m_s = \frac{qc^2(l + 2l_1)}{6l} \tag{2.5}$$

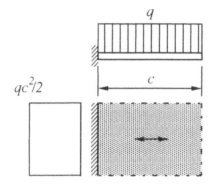

Fig. 2.3.2

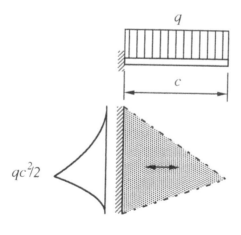

Fig. 2.3.3

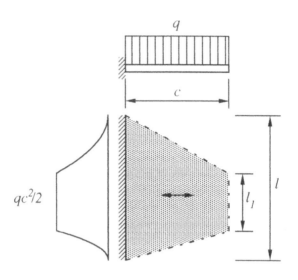

$$\text{Fig. 2.3.4}$$

In an *irregular four-sided one-way element*, Fig. 2.3.5, the sum of average moments is

$$m_f - m_s = \frac{q}{6l}[c_1^2 l_1 + (c_1^2 + c_1 c_2 + c_2^2) l_2 + c_2^2 l_3] \tag{2.6}$$

This formula can be expressed in a general way for slabs with an arbitrary number of sides and with the numbering of lengths following the principles of Fig. 2.3.5:

$$m_f - m_s = \frac{q}{6l}\sum (c_{n-1}^2 + c_{n-1} c_n + c_n^2) l_n \tag{2.7}$$

For the case illustrated in Fig. 2.3.6 the formula is simplified to

$$m_f - m_s = \frac{q}{6}(c_1^2 + c_1 c_2 + c_2^2) \tag{2.8}$$

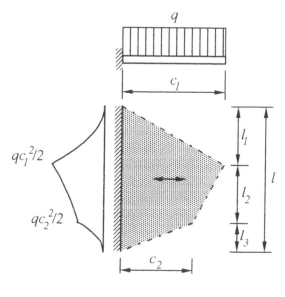

Fig. 2.3.5

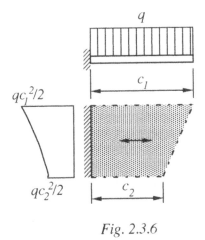

Fig. 2.3.6

2.3.3 Loads with a linear variation in the reinforcement direction

The load is assumed to vary from zero at the support to q_o per unit area a distance c from the support.

In a *rectangular one-way element*, Fig. 2.3.7, the sum of average moments is

$$m_f - m_s = \frac{q_o c^2}{3} \qquad (2.9)$$

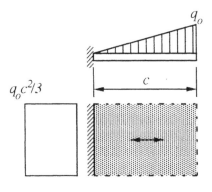

Fig. 2.3.7

In a *triangular one-way element*, Fig. 2.3.8, the sum of average moments is

$$m_f - m_s = \frac{q_o c^2}{12} \qquad (2.10)$$

In a *trapezoidal one-way element*, Fig. 2.3.9, the sum of average moments is

$$m_f - m_s = \frac{q_o c^2 (l + 3l_1)}{12 l} \qquad (2.11)$$

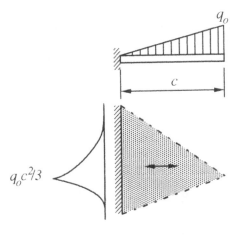

Fig 2.3.8

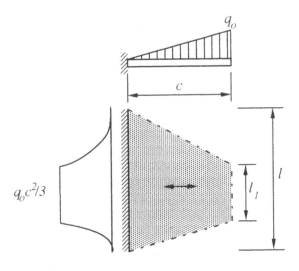

Fig. 2.3.9

2.3.4 Loads with a linear variation at right angles to the reinforcement direction

The load is assumed to vary between zero at the top of the slabs in the figures to q_o at the bottom of the slab, i.e. within a distance l.

In a *rectangular one-way element*, Fig. 2.3.10, the sum of average moments is

$$m_f - m_s = \frac{q_o c^2}{4} \tag{2.12}$$

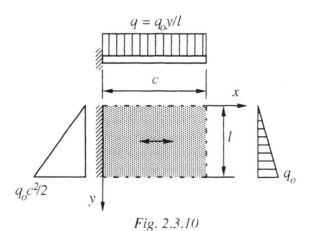

Fig. 2.3.10

In a *triangular one-way element*, Fig. 2.3.11, the sum of average moments is

$$m_f - m_s = \frac{q_o c^2 (l + 2l_1)}{24l} \tag{2.13}$$

In a *trapezoidal one-way element*, Fig. 2.3.12, the sum of average moments is

$$m_f - m_s = \frac{q_o c^2}{24 l^2} \left[3l_1^2 + 6l_2 (2l_1 + l_2) + l_3 (4l - 3l_3) \right] \tag{2.14}$$

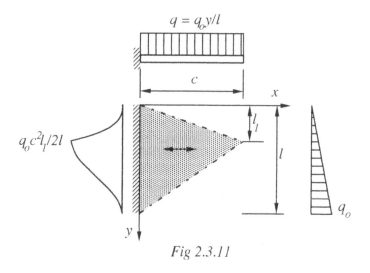

$$q = q_o y/l$$

c

x

$q_o c^2 l_1/2l$

l_1

l

q_o

y

Fig 2.3.11

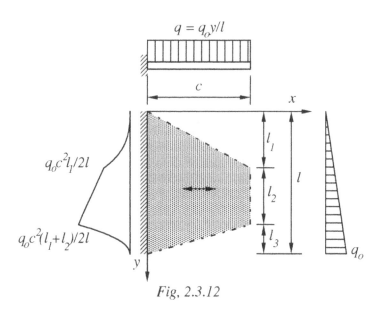

$$q = q_o y/l$$

c

x

$q_o c^2 l_1/2l$

l_1

l

l_2

$q_o c^2(l_1+l_2)/2l$

l_3

q_o

y

Fig. 2.3.12

23

2.3.5 Elements with a shear force along an edge

In some cases the shear force is not zero along the edge of an element. A typical case is where a shear force has a linear intensity variation along an edge according to Fig. 2.3.13. The average moment is then

$$m_f - m_s = \frac{1}{6}[Q_1(2c_1 + c_2) + Q_2(c_1 + 2c_2)] \tag{2.15}$$

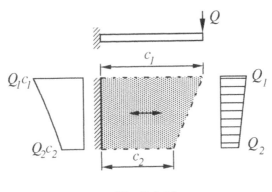

Fig. 2.3.13

2.3.6 Elements with a skew angle between span reinforcement and support

In some slabs it is natural to have different directions for support and span reinforcement. This is the case for triangular slabs and other slabs with non-orthogonal edges. The support reinforcement should normally be arranged at right angles to the support, as this is the most efficient arrangement for taking the support moment and for limiting crack widths. Span reinforcement is often arranged in two orthogonal layers.

The most direct way of treating the case of different directions of support and span reinforcement is through the introduction of a *line (or curve) of zero moment*. On one side of this line the moment is positive and on the other side it is negative. The positive moments are taken by the span reinforcement and the negative moments by the support reinforcement. The load is assumed to be carried in the directions of the reinforcement, that is in different directions on each side of the line of zero moment. We can make a distinction between *span strips* and *support strips*.

Along the line of zero moment shear forces are acting. These shear forces originate from the load on the span strips. The lines of zero moment act as free supports for the span strips. The support strips act as cantilevers, carrying the load on the strips and the shear forces from the span strips.

The shear force in a strip is normally expressed as a force Q per unit width at right angles to the reinforcement direction. Where a span strip is supported on a support strip at a line of zero moment the widths of the cooperating strips are not the same. Using notation according to Fig. 2.3.14, we get the following relation between the shear forces per unit width:

$$Q_2 = \frac{\sin \varphi_1}{\sin \varphi_2} Q_1 \qquad (2.16)$$

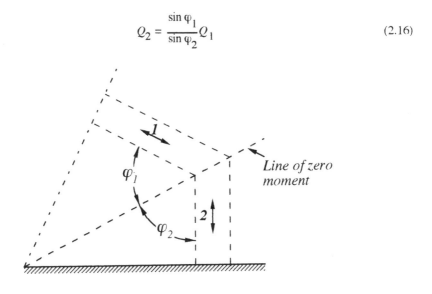

Fig. 2.3.14

In this way it is possible to calculate the positive and negative design moments and their distributions by means of the simple strip method. Examples of such calculations are given in Chapter 6.

This approach is only suitable where the span strips are carrying all the load in one direction. In many cases where the directions of the support and span reinforcements are different the span reinforcement in two directions cooperate in carrying the load on an element. Then the following general approach can be used, where the equilibrium of the whole element is considered, taking into account the moment taken by the span reinforcement in both directions.

The left-hand part in Fig. 2.3.15 shows an element where the span reinforcement is arranged parallel to the x- and y-axes, whereas the support reinforcement is arranged at right angles to the support, which forms an angle φ with the x-axis. The reinforcement in the x- and y-directions corresponds to average moments m_{xf} and m_{yf} and the support reinforcement corresponds to an average moment m_s. The total moments acting on the element are given in the figure.

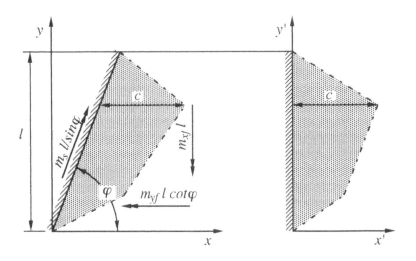

Fig. 2.3.15

The right-hand part of Fig. 2.3.15 shows a corresponding element which is "rectified" so that the support is at right angles to the x-axis. The distances in the x-direction are the same for the two elements. The areas of the elements are the same. Each small area in the left-hand element has a corresponding area of the same size in the right-hand element. The distance at right angles to the support for such an area in the left-hand element is $\sin\varphi$ times the distance at right angles to the support in the right-hand element. Assuming the same load per unit area at the corresponding points in the two elements, the moment with respect to the support for the left-hand element is thus $\sin\varphi$ times the moment in the right-hand element. If we denote the average moment caused by the load in the right-hand element m_0, the total moment caused by the load in the left-hand element is thus $m_0 l/\sin\varphi$.

We can now write the equilibrium equation for the left-hand element, which is the actual element we are interested in:

$$m_{xf} l \sin\varphi + m_{yf} l \cot\varphi \cos\varphi - m_s l / \sin\varphi = m_o l \sin\varphi \tag{2.17}$$

which can be rearranged into

$$m_{xf} + m_{yf} \cot^2\varphi - \frac{m_s}{\sin^2\varphi} = m_o \tag{2.18}$$

In this equation m_0 is the average moment for the right-hand element in Fig. 2.3.15. For the elements treated in equations (2.3)-(2.15) it corresponds to the right-hand side of these equations. It can be seen that these equations correspond to $\varphi = 90°$ in Eq. (2.18).

If the span reinforcement is not orthogonal, but with one reinforcement direction parallel with the x-axis and the other reinforcement direction parallel to the support, the second term in Eq. (2.18) vanishes.

This approach is not a use of one-way elements, as the load is carried in more than one direction. Unlike in the use of one-way elements it is in this case not possible to determine how the load is carried within the element. It is also not possible to determine a theoretical lateral distribution of design moments. An estimate of a suitable distribution of design moments can however be based on the distributions for one-way elements of the same general shape.

2.4 Design moments in one-way elements

2.4.1 General considerations

The strip method gives in principle an infinite number of possible permissible moment distributions. For practical design a solution should be chosen which suits our demands. The main demands are:

1. Suitable behaviour under service conditions.
2. Good reinforcement economy, including simplicity in design and construction.

In discussing moment distributions there are two different types of distribution to take into account, viz. distribution between support and span moments (distribution in the reinforcement direction) and lateral distribution (distribution at right angles to the reinforcement direction). The ratio between support and span moments is discussed in Section 2.9.1.

2.4.2 Lateral distribution of design moments

In the application of the simple strip method average moments in one-way elements are first calculated. In a rigorous analysis using the strip method the moment is not normally constant across the section, but varies due to the varying lengths of the thin one-way strips, and sometimes also due to varying load intensities. The formal moment variation across the section is shown for different cases in Figs 2.3.1-13.

For a rectangular element with a uniform load the moment is constant across the width. In this case the average moment can be directly used for design. In all other cases the strict moment distribution is not uniform, but decreases towards one or both sides. From a practical point of view it is not possible to follow these theoretical distributions in detail, and it is also not necessary, as the behaviour of the slab is not sensitive to limited variations in the lateral reinforcement distribution. On the other hand the choice of an evenly distributed reinforcement corresponding to the average moment may in many cases be too rough an approximation.

A reinforcing bar is usually more active and therefore more beneficial for the behaviour of the slab if it is situated where the curvature of the slab in the direction of the bar is high. Bars which are parallel and close to a support are not very active, as there is practically no curvature in their direction. This fact should be taken into account in the distribution of design moments. It may even be rational to leave parts with a small curvature totally without reinforcement and assume a zero design moment on a certain width of the element. As this would probably not be accepted by some codes this possibility has not been applied in the majority of the examples.

Where the theoretical strict moment distribution is uneven it is generally recommended that one design moment value is chosen for the part where the greatest theoretical moments occur and a smaller design moment is chosen outside this part.

Where an average moment m_{av}, acting across a width l, determines the design moments m_{d1} on width l_1 and the design moment m_{d2} on width $(l-l_1)$, Fig. 2.4.1, the following relation is valid:

$$m_{d1} l_1 + m_{d2} (l - l_1) = m_{av} l \qquad (2.19)$$

With chosen values of the ratios m_{d2}/m_{d1} and l_1/l, the design moment m_{d1} can be calculated from the following formula:

$$m_{d1} = \frac{m_{av}}{\dfrac{l_1}{l} + \dfrac{m_{d2}}{m_{d1}} \left(1 - \dfrac{l_1}{l}\right)} \qquad (2.20)$$

The ratio m_{d2}/m_{d1} is often chosen as 1/2 or 1/3 in order to achieve a simple reinforcement arrangement. In the numerical examples the value 1/2 is often used.

The suitable choice of l_1/l depends on the shape of the element and the load distribution. Proposals for this choice are given in the examples.

Where another distribution is chosen, for example, with three different values of design moments, the same principle may of course be used.

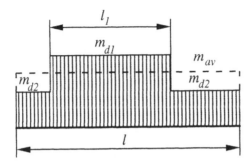

Fig. 2.4.1

2.5 Design moments in corner-supported elements

2.5.1 Corner-supported elements

A corner-supported element is an element which is only supported at one corner. Along all the edges the shear forces and the torsional moments (referred to the reinforcement directions) are zero. In figures the edges are shown as dash-dot lines, indicating zero shear forces.

The load on a corner-supported element has to be carried in two (or more) directions into the supported corner. It has to have active reinforcement in two (or more) directions. It is here assumed that there are only two reinforcement directions, which are usually at right angles to each other. The full load on the element has to be used for the calculation of moments in both directions. This is illustrated by crossing double-headed arrows in the reinforcement directions.

Each reinforcement direction coincides with the direction of one of the edges of the element.

Fig. 2.5.1 shows examples of corner-supported elements with a support at the lower left corner.

Torsional moments exist within a corner-supported element, as it is not possible to carry a load to one point without such moments. Both reinforcement directions cooperate in carrying the torsional moments. A certain amount of reinforcement is required for this purpose in addition to the reinforcement for carrying the bending moments. The total amount of reinforcement required is taken into account in the rules given below, which express the required reinforcement as design bending moments.

With the rules and limitations given below the maximum design moments occur at the edges of the elements, which means that only the edge bending moments need to be calculated as a basis for reinforcement design. Without such rules and limitations higher design

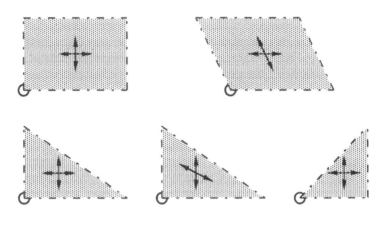

Fig. 2.5.1

moments might occur inside the element, which would complicate the analyses and be uneconomical.

2.5.2 Rectangular elements with uniform loads

The dominating type of corner-supported element in practical design is the rectangular element with a uniform load. This case has therefore been investigated in more detail, leading to detailed design rules. The rules and their background are given in *Strip Method of Design*. Here the rules will be given in a simplified form, suitable for practical design. The rules are on the safe side, sometimes very much so. Minor deviations from the rules for lateral moment distributions may be accepted.

Fig. 2.5.2 illustrates a rectangular corner-supported element with a uniform load q per unit area. The *average bending moments* along the edges have indices x and y for the corresponding reinforcement directions, s for support and f for span (field).The average moments m_{xs} (support moment, negative sign) and m_{xf} (span moment, positive sign) are acting on the element width c_y. The equilibrium of the element with respect to the y-axis demands that

$$m_{xf} - m_{xs} = \frac{qc_x^2}{2} \tag{2.21}$$

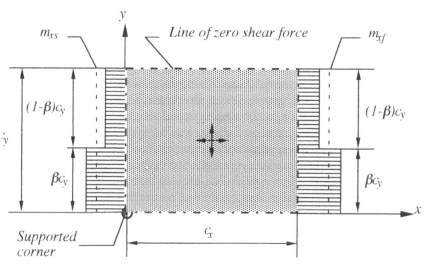

Fig. 2.5.2

These moments are usually distributed on two strips with the widths βc_y and $(1-\beta)c_y$ respectively. The strip with width βc_y, closest to the support, is called the *column strip*, and strip with width $(1-\beta)c_y$ is called the *middle strip*. These terms are chosen in accordance with the normal terms used for flat plates, which are a common application of corner-supported elements.

The distribution of design moments between the two strips has to be such that the numerical sum of moments is higher in the column strip than in the middle strip in order to make moment distribution fulfil the equilibrium conditions within the whole element. The distribution of moments between the strips is defined by the coefficient α:

$$\alpha = \frac{\text{numerical sum of moments in middle strip}}{m_{xf} - m_{xs}} \qquad (2.22)$$

The moment distribution has to fulfil the following conditions

$$\beta \leq 0.5 \qquad (2.23)$$

$$\frac{(1-\beta)}{2} \leq \alpha \leq 0.7 \qquad (2.24)$$

31

In many cases β is chosen as 0.5, and then we have:

$$0.25 \leq \alpha \leq 0.7 \qquad \text{for } \beta = 0.5 \tag{2.25}$$

For the simplest possible arrangement of reinforcement all the support reinforcement is placed within the column strip and the span reinforcement is evenly distributed. This moment distribution is illustrated in Fig. 2.5.3. As in this case there is no support moment in the middle strip the moment within this strip is m_{xf}. Thus we have

$$\alpha = \frac{m_{xf}}{m_{xf} - m_{xs}} = \frac{1}{1 - \dfrac{m_{xs}}{m_{xf}}} \tag{2.26}$$

Combining (2.22) and (2.26) gives

$$\frac{2}{(1 - \beta)} - 1 \geq \frac{-m_{xs}}{m_{xf}} \geq \frac{1}{0.7} - 1 \tag{2.27}$$

or

$$0.43 \leq \frac{-m_{xs}}{m_{xf}} \leq \frac{1 + \beta}{1 - \beta} \tag{2.28}$$

Applying this to $\beta = 0.5$ gives

$$0.43 \leq \frac{-m_{xs}}{m_{xf}} \leq 3 \qquad \text{for } \beta = 0.5 \tag{2.29}$$

This simple moment distribution may be applied for all normally used ratios between support and span moments. For smaller values of β, which means more concentrated support reinforcement, the upper limit of the moment ratio is reduced. It is equal to 2 for $\beta = 1/3$.

Whether this reinforcement arrangement is suitable depends on the demand for crack width control. A reinforcement distribution more in accordance with Fig. 2.5.2 will presumably reduce maximum crack widths, especially for top cracks far from the support.

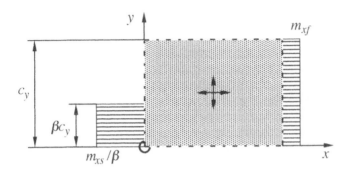

Fig. 2.5.3

2.5.3 Non-rectangular elements with uniform loads and orthogonal reinforcement

Also in non-rectangular elements it is appropriate to divide the element into two strips and distribute the moments between them. It is not possible to give such detailed rules for these cases as for rectangular elements. It is, however, possible to give some general recommendations, which in practice will lead to a safe design.

Fig. 2.5.4 shows triangular corner-supported elements of three different arrangements of reinforcement in the x-direction. With the same definitions as for the rectangular element the following rules are recommended for these slabs:

Case a:

$$\alpha = 0 \qquad 0.3 \leq \beta \leq 0.4 \qquad\qquad (2.30)$$

Case b:

$$0.8 \leq \alpha \leq 1.0 \qquad 0.3 \leq \beta \leq 0.5 \qquad\qquad (2.31)$$

Case c:

$$\alpha = 0 \qquad 0.15 \leq \beta \leq \frac{1}{3} \qquad\qquad (2.32)$$

Most corner-supported elements with orthogonal reinforcement are rectangular or triangular. In cases where other shapes occur they have a shape which is intermediate between rectangular and triangular, and suitable moment distributions may be estimated by means of the recommendations above. It is important to remember that each reinforcement direction must be parallel to an edge of the element.

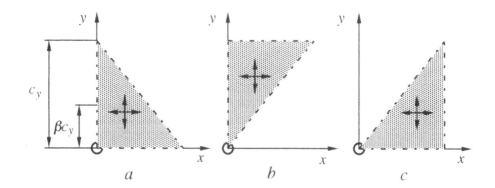

Fig. 2.5.4

2.5.4 Elements with non-orthogonal reinforcement

The determination of the design moments in an element with non-orthogonal reinforcement is based on a so-called *affinity law*. According to this law the design moments are the same as the moments in a "rectified" element with orthogonal reinforcement and the same length and width with respect to the reinforcement in question.

This rule is exemplified in Fig. 2.5.5, which shows a rhomboidal element and the corresponding rectangular elements for the calculation of the design moments in the x- and y-directions. Thus the design moments and the resulting distributions for the reinforcement in the x-direction are determined from the slab in the x_1-y_1-system and for the reinforcement in the y-direction from the slab in the x_2-y_2-system.

2.5.5 Elements with non-uniform loads

The numerical sum of design moments in each direction is calculated with the same formulas as for one-way elements. In the distribution of design moments between column strip and middle strip regard nust be taken of the load distribution. If the column strip in one direction is more heavily loaded than the middle strip a higher portion of the moment should be taken by the column strip than indicated by the rules for uniform load and vice versa. It is not possible to give exact rules to cover all cases, but as the rules for uniform load are quite wide, it ought to be possible to find safe distributions in most cases starting from the rules for uniform load and modifying them with respect to the actual load distribution.

The case of a concentrated load acting on a corner-supported element is treated in Section 2.6.2.

An alternative treatment of corner-supported elements is discussed in Section 2.8.

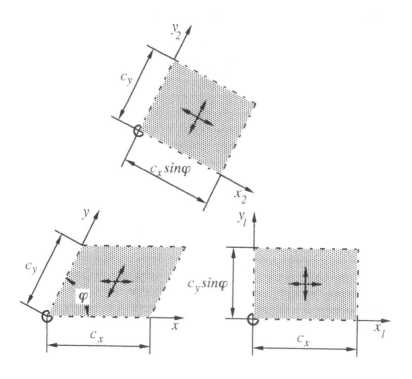

Fig. 2.5.5

2.6 Concentrated loads

2.6.1 One-way elements

A concentrated load is a load which has too high a value per unit area to be taken directly by a one-way strip (or crossing one-way strips) without giving rise to too excessively high local moments. It may be a point load, a line load or a high load on a limited area. The general way of taking care of a concentrated load is by distributing it over a suitable width by means of specially designed distribution reinforcement.

Fig. 2.6.1 shows an example. A concentrated load F acts as a uniform load over a small rectangular area with width b_1 in the x-direction. It is to be carried by a rectangular, simply supported slab. The load is carried on a strip in the y-direction with a width b, chosen to have a moment in the strip which is not too high per unit width. Thus the load F has to be evenly distributed over a strip width b by means of a small strip in the x-direction. The bending moment in this strip is $M_x = F(b - b_1)/8$. This moment should be evenly distributed on a suitable width a, giving a moment

$$m_{xf} = \frac{F(b - b_1)}{8a} \qquad (2.33)$$

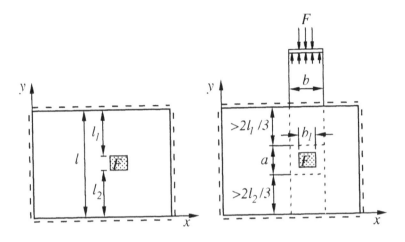

Fig. 2.6.1

It is recommended that the width a should be chosen such that it is approximately centered on the concentrated load and small enough to fulfil the (approximate and somewhat arbitrary) requirements given in the figure. The value of b is chosen so as to get reinforcement in the strip which is well below balanced reinforcement in order to ensure plastic behaviour.

The load distribution on the width b is mainly of importance for the span reinforcement. If the strip is continuous, the support moment (reinforcement) may be distributed over a larger width than b, as the support in itself acts as a load distributor.

In many cases it may be suitable to divide the concentrated load over two strips, one in the x-direction and one in the y-direction, in order to get a reinforcement distribution which is better from the point of view of performance under service conditions. In such a case it is recommended that the load should be divided between the directions approximately in inverse proportion to the ratio between the spans to the fourth power, provided that the support conditions are the same for both strips. If the strips have different support conditions this should be taken into account so that the load taken by a strip with fixed ends can be increased compared to the load taken by a simply supported strip.

When concentrated loads are acting on a slab together with distributed loads special load distribution reinforcement is not necessary if the distributed loads are dominating. This may be considered to be the case if a concentrated load is less than 25% of the sum of distributed loads. Then the concentrated load is simply included in the equilibrium equations for the elements. The design moments may also be redistributed so as to have more reinforcement near the concentrated load.If there are several concentrated loads this simplified procedure may be used even if the sum of the concentrated loads is much higher, and particularly if the concentrated loads are not close to each other.

Applications are shown in the numerical examples in Section 3.3.

2.6.2 Corner-supported elements

Corner-supported elements already have reinforcement in two directions and are thus often able to take care of concentrated loads without any special distribution reinforcement.

The point load F on the corner-supported element in Fig. 2.6.2 gives rise to bending moments $M_{xf} - M_{xs} = Fl_x$. It is generally recommended that the numerical sum of span and support moments should be distributed evenly over the width l_y. Over this width the moment is then $m_{xf} - m_{xs} = Fl_x/l_y$. Some variations from this basic rule are acceptable, e.g. a distribution over a somewhat larger width.

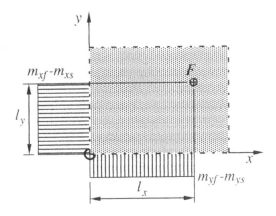

Fig. 2.6.2

Even though the numerical sum of span and support moments has a uniform lateral distribution it is recommended that the span moment should be concentrated in the vicinity of the load and the support moment in the vicinity of the supported corner. Such a distribution is in better agreement with the moment distribution under service conditions and will limit cracking.

When l_y is close to zero the bending moment per unit width becomes too large for the reinforcement in the x-direction according to the expression above. In such cases some extra distribution reinforcement may be necessary, designed according to the general principle for one-way elements.

For applications see Section 9.4.

2.7 Strips

2.7.1 Combining elements to form strips

The elements into which the slab is divided have to be combined in such a way that the equilibrium conditions are fulfilled. These conditions are related to the bending moments and the shear forces at the edges of the elements. Torsional moments are never present at the edges of the elements.

Shear forces are in most cases assumed to be zero at those edges of the elements, which are not supported, but in some applications non-zero shear forces may appear at such an edge.

One-way elements with the same load-bearing direction are often not directly connected to each other, but by means of a constant moment transferred through elements with another load-bearing direction, cf. Fig. 2.2.3. A typical example is shown in Fig. 2.7.1, of a rectangular slab with a uniform load, two simply supported edges and two fixed edges. The positive moments in elements 1 and 3 have to be the same, and this moment is transferred through 2 and 4, which in practice means that the reinforcement is going between 1 and 3 through 2 and 4.

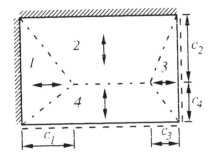

Fig. 2.7.1

The choice of c-values is based on the rules in Section 2.2. It may in practice also be influenced by rules for minimum reinforcement if the optimum reinforcement economy is sought.

In flat slabs corner-supported elements are often directly connected with each other and with rectangular one-way elements into continuous strips as in Fig. 2.7.2. Each such strip acts as a continuous beam, and can be treated as such. The slab in Fig. 2.7.2 thus has a continuous strip in three spans with width w_x in the x-direction and a continuous strip in two spans with width w_y in the y-direction. The design of the slab is based on the analysis of these two strips, taking into account the rules for reinforcement distribution for corner-supported elements. The average design moments in the triangular elements near the corners of the slab are equal to one-third of the corresponding moments in the adjoining rectangular elements.

More detailed rules and numerical examples are given in Chapter 8.

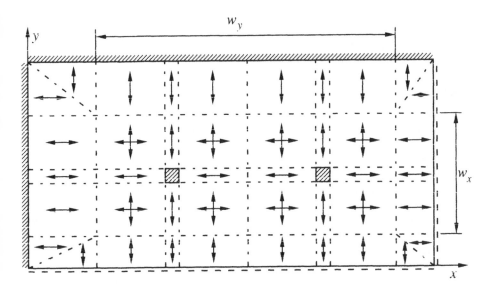

Fig. 2.7.2

2.7.2 Continuous strips with uniform loads

In a continuous strip formed according to Fig. 2.7.2, for example, each part of the strip between two supports, can be treated using normal formulas for beams. The calculation of design moments usually starts with an estimate of suitable support moments. At a continuous support the moment in a strip with a uniform load is normally chosen with respect to what can be expected according to the theory of elasticity, cf. section 2.9.1. After the support moments have been chosen, the distance c_l in Fig. 2.2.2 can be calculated from the formula

$$c_1 = \frac{l}{2} - \frac{m_{s1} - m_{s2}}{ql} = \frac{l}{2} + \frac{-m_{s1} + m_{s2}}{ql} \qquad (2.34)$$

and the span moment from the formula

$$m_f = \frac{qc_1^2}{2} + m_{s1} \qquad (2.35)$$

2.8 Support bands

2.8.1 General

A support band is a band of reinforcement in one direction, acting as a support for strips in another direction. By means of support bands it is possible to make direct use of the general principles of the simple strip method for all types of slab. It is the most general method, and the method which has to be used in cases where other approaches cannot be applied.

A reinforcement band of course has a certain width. In *Strip Method of Design* it has been demonstrated how reinforcement bands may be used in a strictly correct way, taking into account the widths of the bands. In order to simplify the analyses the following approximation will be accepted here for the calculation of the design moment in the support band:

In the numerical analysis the support band is assumed to have zero width.

The reinforcement for the moment in the support band is distributed over a certain width, which is limited by rules given in Section 2.8.3. If these rules are followed the safety at the ultimate limit state can be estimated to be sufficient in spite of the approximation.

2.8.2 Comparison with corner-supported elements

A rectangular corner-supported element with a uniform load can be analysed by means of simple strips supported on support bands along the coordinate axes, which are in their turn supported at the corner.

Fig. 2.8.1 a) shows an element where half the load is assumed to be carried in each direction. This gives evenly distributed moments $m_{xf} - m_{xs} = qc_x^2/4$, and a reaction force on the support band along the x-axis equal to $qc_y/2$. This reaction force on the support band gives concentrated moments $M_{xf} - M_{xs} = qc_y c_x^2/4$, corresponding to an average moment $qc_x^2/4$ on the whole width c_y. Thus half the total moment $qc_x^2/2$ is evenly distributed and the other half is concentrated in the assumed support band of zero width.

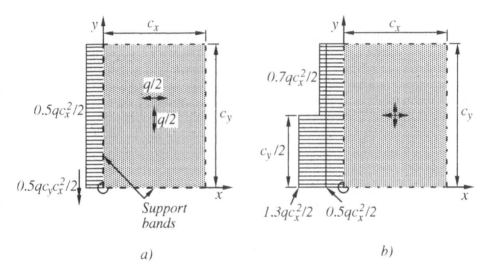

Fig. 2.8.1

The moments calculated from an assumption of support bands in Fig. 2.8.1 a) can be compared to the moment distribution according to Fig. 2.8.1 b), which is acceptable according to condition (2.25). From this comparison it can be concluded that in this case it is acceptable to distribute concentrated moments from an assumed support band of zero width over half the width of the element and that the solution is still on the safe side. This conclusion is drawn for a rectangular element with a uniform load, but indirectly it can also be concluded that moments from an assumed support band of zero width can always be distributed over a certain width of the element. The most suitable choice of this width depends on the load distribution and the shape of the element. If more of the load is acting near the support band a smaller width should be chosen. With some caution it is not difficult to choose a width which is safe. Detailed recommendations are given below.

It may also be noted that according to conditions (2.23) and (2.24) the moment from the support band may be evenly distributed over any arbitrary width between zero and $l_y/2$.

2.8.3 Application rules

The moments in the support band are distributed over a certain width to give the design moments for the reinforcement. The width of the reinforcement band has to be limited so it can accomodate the moments which are concentrated in a band of zero width. It is not possible to establish general rules for the maximum acceptable width of the reinforcement band based on a rigorous solution according to the lower bound theorem of the theory of plasticity. The following recommendations are based on the comparison above with a corner-supported element and on estimates. They are intended for situations where the load on the slab is uniform.

The width of the reinforcement band is based on a comparison with the average width of the elements which are supported by the support band. This average width is denoted b_a. The width of the *support* reinforcement in the band may be equal to b_a provided that the band is *supported over its whole width* by a support which is nearly at right angles to the band. Support reinforcement over columns may be distributed over a width of about $0.5b_a$. Span reinforcement may be distributed over a width between $0.5b_a$ and b_a, depending on the importance of the band for the static behaviour of the slab. The more important the band is, the narrower the width of the reinforcement band. The reinforcement should if possible be distributed on both sides of the theoretical support band in proportion to the loads from the two sides.

Where a support band has a support with a strong force concentration certain rules have to be followed in order to prevent local failure.

For the case in Fig. 2.8.2 some minimum top reinforcement is required at the ends of the band to prevent radial cracks from the support point. This reinforcement is placed at right angles to the band and is designed for a bending moment $m_\perp$, determined from the following condition:

$$\left| m_\perp \cdot m_b \right| \geq \left(\frac{2R}{\pi} \times \frac{\varphi_1}{\varphi_1 + \varphi_2} \right)^2 \tag{2.36}$$

where R is the support reaction from the support band, and

m_b is the numerical sum of the span and support moments which are taken by the reinforcement in the reinforcing band at the support where R is acting.

The corresponding reinforcement should be present from the support to a distance equal to one-third of the distance to the point of maximum moment in the support band and be well anchored. It does not need to be additional reinforcement, as the reinforcement arranged for other reasons may cover this need.

A check according to these rules must be made for both ends of the reinforcement band.

Where the band is supported on a column, a check should be made that the following relation is fulfilled

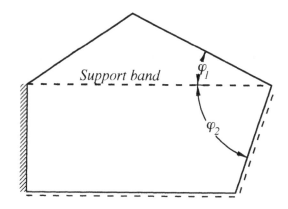

Fig. 2.8.2

$$\sqrt{(m_{xf}-m_{xs})\,(m_{yf}-m_{ys})} \geq \frac{R}{2\pi}\left[1-1.3\,(\frac{qA}{R})^{1/3}\right] \tag{2.37}$$

where the span and support moments correspond to the reinforcement in the vicinity of the column. If the span moments are different on the different sides of the column the average value is used. R is the reaction force at the column, A is the support area at the column and q is the load per unit area in the vicinity of the column.

It is not normally necessary to perform this check, as the recommendations for the design are intended to fulfil this relation automatically.

In addition to the concentrated reinforcement in a band some extra reinforcement may be needed for crack control.

Although it cannot be strictly proven by means of the lower bound theorem that the above recommendations are always on the safe side the design can be considered to be safe. In any case it is always safer than a design based on the yield line theory: this can be checked by applying the yield line theory to slabs designed according to these recommendations.

2.9 Ratios between moments

2.9.1 Ratio between support and span moments in the same direction

The recommendations below follow the general guidelines in Section 1.5.

The strip method does not in itself give ratios between support and span moments, as the equilibrium equations can be fulfilled independently of this ratio. From the point of view of safety at the ultimate limit state this ratio is unimportant. The ratio is, however, of importance for behaviour under service conditions and for reinforcement economy. These factors should therefore be taken into account in the choice of the ratios between support and span moments.

It can be shown that the best choice of the ratio between the numerical values of support and span moments in an element or strip is usually approximately equal to the ratio according to the theory of elasticity or somewhat higher. The ratio may, however, be chosen within rather wide limits without any effect on safety and with only a very limited effect on deformations under service conditions. It mainly has an influence on the width of cracks. With a higher ratio the crack widths above the support are somewhat decreased whereas the crack widths in the span are somewhat increased. If the intention is to limit the crack width on the upper side of the slab a large ratio between span and support moments should therefore be chosen.

Where the load is uniform it is generally recommended that, for continuous strips and for strips with fixed supports, a ratio between the numerical values of support and span moments should be chosen around 2, say between 1.5 and 3.0. For triangular elements, such as at the short edge of a rectangular slab, higher values may be used. Where a strip is continuous over a support the average of the span moments on both sides of the support is used for the calculation of moment ratio.

So for a slab such as that in Fig. 2.7.1 the ratio c_2/c_4 is chosen between 1.6 and 2.0, corresponding to a moment ratio of between 1.56 and 3.0, whereas c_1/c_3 may be chosen between 1.7 and 2.2, corresponding to a moment ratio between 1.9 and 3.8.

Where it is possible to estimate the moments according to the theory of elasticity this should be done as a basis for the determination of the main moments. This is the case, for instance, where the advanced strip method is used for flat slabs. The support moment at a support where the slab is continuous can be taken to be approximately equal to the average of the moments corresponding to fixed edges for the spans on the two sides of the support.

For the irregular flat slabs discussed in Chapter 10 a special approach for the determination of support moments has been introduced in order to find support moments which are also in approximate agreement with the theory of elasticity in these complex cases.

2.9.2 Moments in different directions

In some applications there is no real choice of moment distribution in different directions. For example, this is the case for the flat slab in Fig. 2.7.2, where the distribution is given by the analysis of the strips. This is due to the fact that within the major part of the slab the load is carried by corner-supported elements, which carry the whole load in both directions, and where the load is thus not distributed between the directions.

In slabs where the load is carried by one-way elements, e.g. the slab in Fig. 2.7.1, it is possible to make a choice of the direction of load-bearing reinforcement within those parts of the slab where the elements with different directions meet. Thus, for the slab in Fig. 2.7.1, it is possible to increase or decrease c_1 and c_3, leading to increased or decreased moments in elements *1* and *3* and corresponding decreases or increases of the moments in the opposite direction. The choice of directions of the lines of zero shear force starting at the corners has been discussed in Section 2.2.

2.10 Length and anchorage of reinforcing bars

2.10.1 One-way elements

In principle, the length of reinforcing bars is determined from the curve of bending moments which shows the variation of bending moments in the direction of the strip. This curve is easy to determine for rectangular elements with a uniform load, but in most other cases the curve is not well defined. Then the necessary length of reinforcing bars has to be determined by means of some approximate rule, which should be on the safe side.

For *one-way elements with a uniform load* the following rules are recommended:

1. One half of the bottom reinforcement is taken to the support. Close to a corner of a slab, however, *all* the bottom reinforcement is taken to the support.

2. The other half of the bottom reinforcement is taken to a distance from the support equal to

$$c\left[1 - \left(2\left(1 - \frac{m_s}{m_f}\right)\right)^{-1/3}\right] - \Delta l \tag{2.38}$$

3. One half of the top reinforcement is taken to a distance from the support equal to

$$c\left[1 - \left(1 - \frac{m_s}{2m_f}\right)^{1/2}\left(1 - \frac{m_s}{m_f}\right)^{-1/2}\right] + \Delta l \tag{2.39}$$

4. The rest of the top reinforcement is taken to a distance from the support equal to

$$c \left[1 - \left(1 - \frac{m_s}{m_f} \right)^{-1/2} \right] + \Delta l \tag{2.40}$$

In these formulas c is the length of the element, shown in Figs 2.2.2–5, and Δl is an additional length of the reinforcing bars for anchorage behind the point where the moment curve shows that it is not needed any more. This additional length, which depends on the slab depth and on the diameter of the reinforcement, is given in many national codes. If this is not the case, it is recommended that a value is used equal to the depth of the slab.

It is not possible to give exact rules covering all possible situations with respect to load distributions. In cases other than those treated above it can only be recommended that reinforcement lengths are chosen which are judged to be on the safe side.

As an example, the application of the above formulas to the reinforcement in the x-direction will be shown for the slab in Example 3.2. The value of Δl is assumed to 0.15 m. The moment values are $m_{xs} = -7.20$, $m_{xf} = 3.92$, which gives $m_{xs}/m_{xf} = -1.837$.

For element 1 in Fig. 3.1.3 we have $c = 2.6$ m. In this element half the bottom reinforcement is taken to the support and the other half, according to Eq. (2.38), to a distance from the support equal to

$$2.6 \left[1 - (2 \times 2.837)^{-1/3} \right] - 0.15 = 0.99 \text{ m} \tag{2.41}$$

One half of the top reinforcement is, according to Eq. (2.39), taken to a distance from the support equal to

$$2.6 \left(1 - 1.918^{1/2} \times 2.837^{-1/2} \right) + 0.15 = 0.61 \text{ m} \tag{2.42}$$

The rest of the top reinforcement is, according to Eq. (2.40), taken to a distance from the support equal to

$$2.6 \left(1 - 2.837^{-1/2} \right) + 0.15 = 1.21 \text{ m} \tag{2.43}$$

For element 3, with $c = 1.4$ m, the support moment is zero and there is no support reinforcement. Half the bottom reinforcement is taken to the support and the other half to a distance from the support equal to

$$1.4 \left(1 - 2^{-1/3} \right) - 0.15 = 0.14 \text{ m} \tag{2.44}$$

This is a small distance and from an economic point of view all the reinforcement may as well be taken to the support. It is generally recommended that all bottom reinforcement is taken to the support in a slab which is simply supported.

The length of reinforcing bars will not be calculated in the numerical examples, as the calculation is simple and sometimes other rules have to be followed according to codes.

2.10.2 Corner-supported elements

The distribution of design moments within a corner-supported element is extremely complex, due to the fact that torsional moments play an important part in carrying the load to a corner of the element. These torsional moments have to be taken into account in the design of reinforcement, which is formally done as an addition to the design bending moments which the section have to resist. It is hardly possible to find solutions which cover all situations in a theoretically correct way. The rules given below are basically on the safe side at the same time as giving approximately the same results as common design methods for flat slabs.

In *all corner-supported elements* all bottom reinforcement should be taken all the way to the edges of the element.

In *rectangular corner-supported elements with a uniform load* half the top reinforcement should be taken to a distance from the support line equal to

$$\gamma c + \Delta l \tag{2.45}$$

The rest of the top reinforcement should be taken to a distance from the support line equal to

$$\frac{2}{3}\gamma c + \Delta l \tag{2.46}$$

In these formulas Δl has the same meaning as in 2.10.1, γ depends on the ratio between support and span moments according to Fig. 2.10.1 and c is the length of the element in the direction of the reinforcement.

2.10.3 Anchorage at free edges

Where a reinforcing bar ends at a free edge it must be satisfactorily anchored. In many cases a deformed bar may just end at the edge, but where the force in the bar is expected to be high close to the edge, the bar should be provided with an extra strong anchorage, e.g. according to Fig. 2.9.2. This will occur where there is a short distance from a line of maximum moment (line of zero shear force) to the edge or where large torsional moments are acting in the slab, which is the case in the vicinity of point supports or where point loads are acting close to a free edge, for example. Large torsional moments also act in slabs with two adjacent free edges.

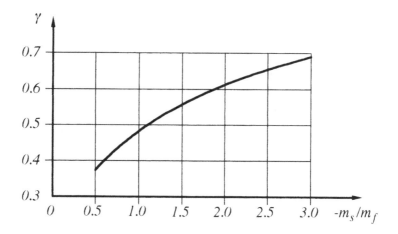

Fig. 2.10.1

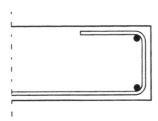

Fig. 2.10.2

2.11 Support reactions

In one-way elements the load is carried in the load-bearing direction into the support. It is thus simple to determine the support reaction and its distribution. With a uniform load on an element the distribution corresponds to the shape of the element. The accurate distribution of the reaction should be taken into account in the design of supporting structures like beams and support bands.

Where the support is not parallel to one of the two span reinforcement directions, as in Fig. 2.3.15, it is not possible to use one-way elements and therefore not possible to determine an accurate distribution of support reactions. In such a case the load is assumed to be carried at right angles to the support (the shortest distance to the support). This assumption gives the correct total support reaction and the resulting moments in a supporting beam or band will in all practical cases be on the safe side. The maximum shear force in a supporting beam or band may be slightly on the unsafe side.

For corner-supported elements the whole load on the element is theoretically acting as a support reaction at the supported corner. In reality it is of course distributed in some way over the area of the support. The resultant to the support forces is determined by the theoretical forces from the elements acting upon the support.

CHAPTER 3

Rectangular slabs with all sides supported

3.1 Uniform loads

3.1.1 Simply supported slabs

Example 3.1

The slab in Fig. 3.1.1 is simply supported with a uniform load of 9 kN/m^2. The lines of zero shear force are chosen as shown in the figure in accordance with the rules given in Section 2.2. The average span moment in the x-direction is, Eq. (2.4):

$$m_{xf} = \frac{9 \times 2.0^2}{6} = 6.0 \tag{3.1}$$

The average span moment in the y-direction is, Eq. (2.5):

$$m_{yf} = \frac{9 \times 2.2^2 (6.5 + 2 \times 2.5)}{6 \times 6.5} = 12.84 \tag{3.2}$$

The lateral distribution of design moments follows the general recommendations in Section 2.4.2. Thus, for example, we may choose to have moments in the side strips which are half those in the central strip (and with distances between bars twice as large). If we choose

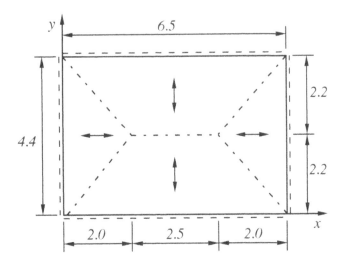

Fig. 3.1.1

the widths of the side strips to 1.1 m in both directions, we get in the central strips, Eq (2.20):

$$m_x = \frac{6.0}{\frac{2.2}{4.4} + 0.5\left(1 - \frac{2.2}{4.4}\right)} = 8.0 \tag{3.3}$$

$$m_y = \frac{12.84}{\frac{4.3}{6.5} + 0.5\left(1 - \frac{4.3}{6.5}\right)} = 15.46 \tag{3.4}$$

This moment distribution is illustrated in Fig. 3.1.2.

3.1.2 Fixed and simple supports

Example 3.2

The slab in Fig. 3.1.3 has the same size and load (9 kN/m^2) as the previous slab, but the upper and left-hand edges are fixed and there are negative moments along these edges. The numerical sum of moments in these elements is thus greater than in the elements at the sim-

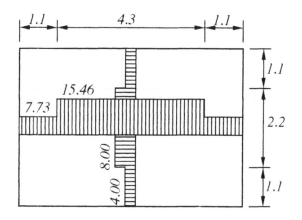

Fig. 3.1.2

ply supported edges. This is achieved by choosing suitable sizes of the elements, see Section 2.9.1. With the choice of size as shown in Fig 3.1.3 for elements 1 and 3 we get from Eq. (2.4)

$$m_{xf} = \frac{9 \times 1.4^2}{6} = 2.94 \tag{3.5}$$

$$m_{xf} - m_{xs} = \frac{9 \times 2.6^2}{6} = 10.14 \tag{3.6}$$

$$m_{xs} = -10.14 + 2.94 = -7.20 \tag{3.7}$$

For elements 4 and 2 we get from Eq. (2.5)

$$m_{yf} = \frac{9 \times 1.6^2 (6.5 + 2 \times 2.5)}{6 \times 6.5} = 6.79 \tag{3.8}$$

$$m_{yf} - m_{ys} = \frac{9 \times 2.8^2 (6.5 + 2 \times 2.5)}{6 \times 6.5} = 20.81 \tag{3.9}$$

$$m_{ys} = -20.81 + 6.79 = -14.02 \tag{3.10}$$

53

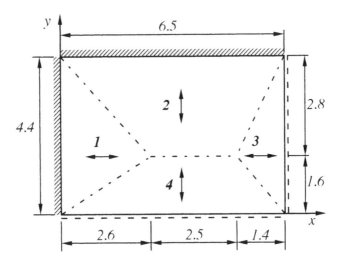

Fig. 3.1.3

The numerical ratios between the support and span moments are 2.45 in the x-direction and 2.07 in the y-direction. These values are acceptable (see Section 2.9.1), but if, for example, we wish to have a slightly smaller ratio for the y-direction we can change the dividing line (e.g. with c-values of 2.7 and 1.7 instead of 2.8 and 1.6) and repeat the analysis. We then find $m_{yf} = 7.76$ and $m_{ys} = -11.59$ and the ratio 1.49. This ratio is evidently sensitive to the choice of the position of the line of zero shear force.

If we choose the same type of moment distribution as in the previous example we find the distribution of design moments according to Fig. 3.1.4.

The determination of the lengths of reinforcing bars for the slab in this example is demonstrated in Section 2.10.1.

Example 3.3

The slab in Fig. 3.1.5 has a load of 11 kN/m². The left-hand edge is fixed, whereas the right-hand edge is elastically restrained so that the support moment is lower. This is taken into account in the analysis by assuming a non-symmetrical pattern of lines of zero shear force, as shown in the figure.

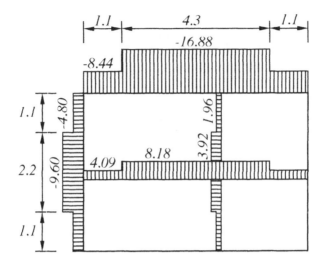

Fig. 3.1.4

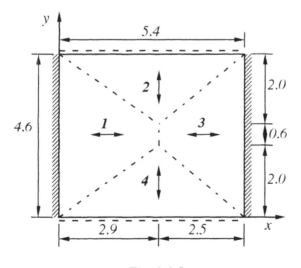

Fig. 3.1.5

Using Eq. (2.1) we get for element 1

$$m_{xf} - m_{xs1} = \frac{11 \times 2.9^2 (4.6 + 2 \times 0.6)}{6 \times 4.6} = 19.44 \tag{3.11}$$

and for element 3

$$m_{xf} - m_{xs2} = \frac{11 \times 2.5^2 (4.6 + 2 \times 0.6)}{6 \times 4.6} = 14.45 \tag{3.12}$$

We can, for instance choose m_{xf}=7.00, which gives m_{xs1}=−12.44; m_{xs2}=−7.45. From Eq. (2.4) we get for elements 2 and 4

$$m_{yf} = \frac{11 \times 2.0^2}{6} = 7.33 \tag{3.13}$$

In calculating design moments according to Eq. (2.20) the widths of the side strips are chosen equal to 1.0 m in the x-direction and 1.5 m in the y-direction. The reason why a higher value has been chosen for the y-direction is that reinforcement along a fixed edge is rather inefficient in taking up stresses under normal loads, as the slab has very limited deflections in these regions. The resulting distribution of design moments is shown in Fig. 3.1.6.

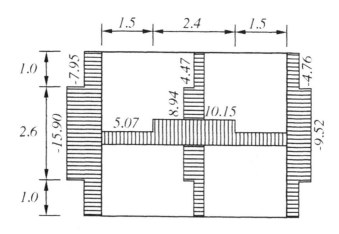

Fig. 3.1.6

3.2 Triangular loads

Example 3.4

The slab in Fig. 3.2.1 is simply supported along all edges. It is a vertical slab acted upon by water pressure, which is zero at the upper edge and is assumed to increase by 10 kN/m² per m depth. (This value is an approximation used to make it easier to follow the analysis. A more correct value is about 10.2.) For reasons of symmetry $c_1 = c_3 = (4.8 - l_1)/2$. The value of $c_4 = 3.4 - c_2$ is determined by the condition that the span moments in elements 2 and 4 must be the same.

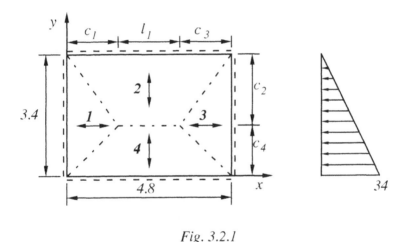

Fig. 3.2.1

For the different elements we have the following expressions for the moments, calculated from Eq. (2.13) for elements 1 and 3, from (2.11) for element 2 and from a combination of (2.5) and (2.11) for element 4 (evenly distributed load of 34 minus a triangular load with the value $10c_4$ at the upper line of zero shear force):

$$m_{xf} = \frac{34c_1^2(3.4 + 2c_2)}{24 \times 3.4} \qquad (3.14)$$

$$m_{yf2} = \frac{10c_2 \times c_2^2(4.8 + 3l_1)}{12 \times 4.8} \qquad (3.15)$$

$$m_{yf4} = \frac{34c_4^2(4.8 + 2l_1)}{6 \times 4.8} - \frac{10c_4 \times c_4^2(4.8 + 3l_1)}{12 \times 4.8} \qquad (3.16)$$

To directly use the condition $m_{yf2} = m_{yf4}$ leads to an equation of the third degree. It is in practice easier to use iteration, trying different values until the condition is fulfilled. Using $l_1 = 1.8$ m it is found that $c_4 = 1.39$ m, and we get the following result for average moments:

$$m_{xf} = 6.96; \quad m_{yf} = 14.40$$

The ratio between the moments in the x- and y-directions should also be checked in order to achieve a good reinforcement economy and satisfactory behaviour under service conditions. In this case this ratio may seem suitable with respect to the spans in the x- and y-directions. The ratio may be changed by means of another choice of the value of l_1. However, neither economy nor serviceability are much influenced by small changes of this type.

With the values above, the distribution of design moments can be chosen in accordance with Fig 3.2.2. The distribution of reinforcement in the x-direction is chosen with less reinforcement in the upper part of the slab, where the load is smaller.

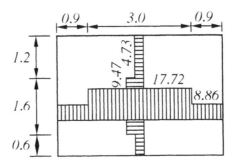

Fig. 3.2.2

Example 3.5

Fig. 3.2.3 shows a slab with all edges fixed, which is acted upon by a load which is 18 kN/m² at the top and increases by 5 kN/m² per metre to 39 kN/m² at the bottom. A choice of lines of zero shear force is shown. For all parts of the slab a combination of uniform load and triangular load has to be used.

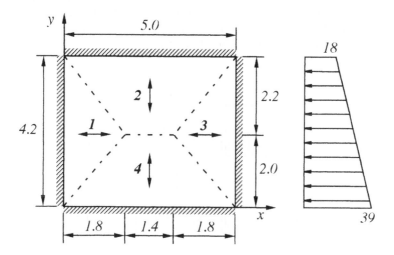

Fig. 3.2.3

For element 1 (identical to element 3) Eq. (2.4) is used with $q = 18$ and Eq. (2.13) is used with $q = 21$:

$$m_{xf} - m_{xs1} = \frac{18 \times 1.8^2}{6} + \frac{21 \times 1.8^2 (4.2 + 2 \times 2.2)}{24 \times 4.2} = 15.53 \qquad (3.17)$$

For element 2 Eq. (2.5) is used with $q = 18$ and Eq. (2.11) is used with $q = 11$:

$$m_{yf} - m_{ys2} = \frac{18 \times 2.2^2 (5.0 + 2 \times 1.4)}{6 \times 5.0} + \frac{11 \times 2.2^2 (5.0 + 3 \times 1.4)}{12 \times 5.0} = 30.81 \qquad (3.18)$$

For element 4, Eq. (2.5) is used with $q = 39$ and Eq. (2.11) is used with $q = -10$:

$$m_{yf} - m_{ys4} = \frac{39 \times 2.0^2 (5.0 + 2 \times 1.4)}{6 \times 5.0} - \frac{10 \times 2.0^2 (5.0 + 3 \times 1.4)}{12 \times 5.0} = 34.43 \qquad (3.19)$$

We can now choose values of m_{xf} and m_{yf}, which give suitable ratios between support and span moments. The following values may be chosen, for example:

$$m_{xf} = 4.53; \; m_{xs} = 11.00; \; m_{yf} = 10.40; \; m_{ys2} = -20.41; \; m_{ys4} = -24.03$$

With these values and side strips in both directions equal to 1.0 m we get the distribution of design moments shown in Fig. 3.2.4. If the ratios between the different moments are not regarded as suitable, either the choice of span moments can be changed, or the pattern of lines of zero moments can be adjusted and the calculation repeated.

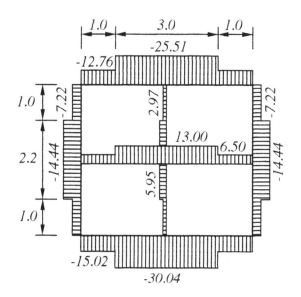

Fig. 3.2.4

Example 3.6

The wall in Fig. 3.2.5 is acted upon by a water pressure with a water level 1.0 m below the upper edge. The pressure is triangular with an assumed increase of 10 kN/m^2 per metre depth. The loading cases on the elements do not directly correspond with those treated in Section 2.3. It is necessary to combine several of these cases in order to calculate the average moments. The combination of loading cases for the calculation of moments in elements 1, 2 and 3 is indicated in Fig. 3.2.5.

In element 1 (and 3) this gives the following evaluation:

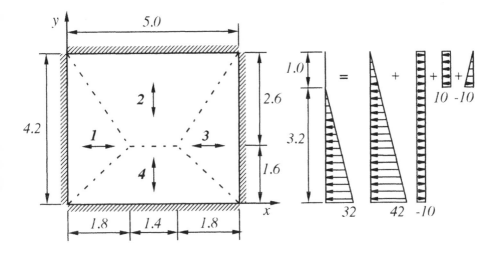

Fig. 3.2.5

$$m_{xf} - m_{xs} = \frac{42 \times 1.8^2 (4.2 + 2 \times 2.6)}{24 \times 4.2} - \frac{10 \times 1.8^2}{6} +$$

$$+ \frac{1.0}{4.2} \times \frac{10 \times 0.69^2}{6} - \frac{1.0 \times 10 \times 0.69^2 \times 3 \times 1.0}{4.2 \times 24 \times 1.0} = 7.34$$

(3.20)

The first term is the influence of a triangular load on the whole element according to Eq. (2.13) and the second term is the influence of a negative uniform load on the whole element according to Eq. (2.4). The third and fourth terms show the influence of loads on only the upper part of the element, down to a depth of 1.0 m, where the height is 0.69 m. The average moment on the upper part is distributed on the whole element width 4.2 m through multiplication by 1.0/4.2.

For element 2 we get

$$m_{yf} - m_{ys2} = \frac{26 \times 2.6^2 (5.0 + 3 \times 1.4)}{12 \times 5.0} - \frac{10 \times 2.6^2 (5.0 + 2 \times 1.4)}{6 \times 5.0} +$$

$$+ \frac{10 \times 1.0^2 (5.0 + 2 \times 3.62)}{6 \times 5.0} - \frac{10 \times 1.0^2 (5.0 + 3 \times 3.62)}{12 \times 5.0} = 10.81$$

(3.21)

The first term is the influence of a triangular load on the whole element according to Eq. (2.11) and the second term is the influence of a negative uniform load on the whole element according to Eq. (2.5). The third and fourth terms give the influence of a uniform load and a negative triangular load on only the upper 1.0 m of the element. The width of the element at that level is 3.62 m.

For element 4 we get

$$m_{yf} - m_{ys4} = \frac{32 \times 1.6^2 (5.0 + 2 \times 1.4)}{6 \times 5.0} - \frac{16 \times 1.6^2 (5.0 + 3 \times 1.4)}{12 \times 5.0} = 15.02 \qquad (3.22)$$

The first term is the influence of a uniform load according to Eq. (2.5) and the second term is the influence of a negative triangular load according to Eq. (2.11).

The following moments are chosen in order to achieve suitable ratios between support and span moments:

$$m_{xf} = 2.00; \; m_{xs} = -5.34; \; m_{yf} = 4.00; \; m_{ys2} = -6.81; \; m_{ys4} = -11.02$$

A suitable distribution of design moments is proposed in Fig. 3.2.6. The reinforcement in the x-direction is not symmetrical, but is concentrated downwards because of the very non-symmetrical load distribution.

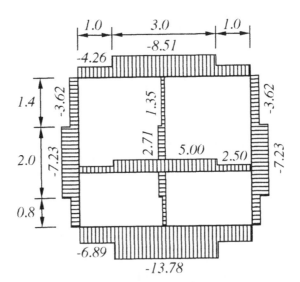

Fig. 3.2.6

3.3 Concentrated loads

3.3.1 General

A concentrated load is a load which acts only on a relatively small part of a slab. It may be a point load, a line load or a load with a high intensity on a small area.

A concentrated load generally acts together with a distributed load. The way that the concentrated load is taken into account in design depends on the relative importance of this load compared to the distributed load. If the concentrated load is important it must be separately taken into account in the design. If the distributed load is dominant the influence of the concentrated load can be included at the same time as the distributed load. It is suggested that a separate calculation is not needed in cases where the concentrated load is less than about 25% of the sum of distributed loads. If two or more concentrated loads are spread over the slab this figure may be increased.

3.3.2 A concentrated load alone

Example 3.7

The slab in Fig. 3.3.1 is acted upon by a point load of 150 kN. It should be designed to carry the whole load in the direction of the shortest span, i.e. in the y-direction. This is the least expensive way to carry this load, but if only the resulting reinforcement is used, the slab will not behave well in the service state. This example is only shown in order to demonstrate the principle used in design for concentrated loads. The next example shows how a concentrated load can be divided between strips in two directions in order to achieve a reinforcement which is better for the behaviour in the service state.

As shown in the figure the load is assumed to be carried by a strip in the y-direction with a width of 2.0 m. The load is distributed over this width by means of reinforcement in the x-direction in a short strip with width 1.5 m. The design moment in this strip is calculated by means of Eq. (2.33):

$$m_{xf} = \frac{150 \times 2.0}{8 \times 1.5} = 25.0 \tag{3.23}$$

The two elements 2 and 4 meet at the point load. Strictly, one part F_4 of the point load is carried by element 2 and the remaining part by element 4. We get the following relations:

$$m_{yf} - m_{ys} = \frac{F_4 \times 2.2}{2.0} \tag{3.24}$$

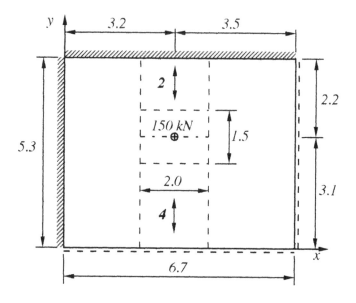

Fig. 3.3.1

$$m_{yf} = \frac{(150 - F_4) \times 3.1}{2.0} \tag{3.25}$$

A suitable choice is $F_4 = 120$, which gives $m_{yf} = 46.5$; $m_{ys} = -85.5$. The resulting distribution of design moments is shown in Fig. 3.3.2.

The distribution reinforcement in the x-direction is mainly intended for the distribution of the bending moment over the width of the span reinforcement in the y-direction. The support reinforcement can be active over a larger width, as the support acts as load distributor and forces the reinforcement to cooperate. The support moment can therefore be distributed over a larger width than the width of 2.0 m assumed for the design. This has been indicated by a dashed line in Fig. 3.3.2.

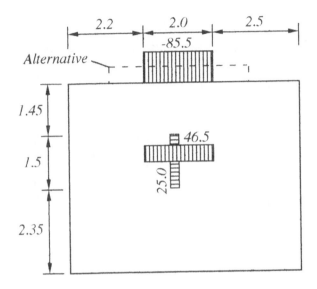

Fig 3.3.2

Example 3.8

This example is intended to illustrate both how a concentrated load is distributed on strips in two directions and how to treat a concentrated load in the form of a high load intensity on a small area. A total load of 150 kN is uniformly distributed on an area of 1.2 m ×0.8 m. The load is distributed on one strip in the x-direction with width 2.0 m, and one strip in the y-direction with width 2.5 m.

When the load is distributed in the two directions it is satisfactory to divide the load approximately inversely to the ratio of spans to the fourth power. In this case we will have 40 kN in the x-direction and 110 kN in the y-direction.

For each direction we can assume a line of zero shear force (maximum moment) at right angles to the load-bearing direction, placed inside the loaded area. The coordinates of these lines are denoted x_1 and y_1.

For the strip in the x-direction we get a moment m_{yf} for load distribution from Eq. (2.33):

$$m_{yf} = \frac{40\,(2.0 - 0.8)}{8 \times 2.5} = 2.4 \tag{3.26}$$

65

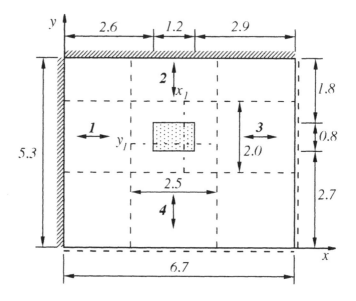

Fig. 3.3.3

For the moments in the two elements 1 and 3 forming the strip we have:

$$m_{xf} - m_{xs1} = \frac{40(x_1 - 2.6)}{1.2 \times 2.0} \times \frac{(x_1 + 2.6)}{2} \tag{3.27}$$

$$m_{xf} = \frac{40 \cdot (3.8 - x_1)}{1.2 \times 2.0} \times (2.9 + \frac{3.8 - x_1}{2}) \tag{3.28}$$

Different values of x_1 are tried in these formulas until a suitable ratio between support and span moments is found. Thus for $x_1 = 3.5$ m we find $m_{xf} = 15.3$; $m_{xs1} = -30.5$.

For the strip in the y-direction we get a moment m_{yf} for load distribution:

$$m_{xf} = \frac{110(2.5 - 1.2)}{8 \times 2.0} = 8.94 \tag{3.29}$$

For the moments in the two elements 2 and 4 forming the strip we have

$$m_{yf} - m_{ys2} = \frac{110\,(3.5 - y_1)}{0.8 \times 2.5} \times (1.8 + \frac{3.5 - y_1}{2}) \tag{3.30}$$

$$m_{yf} = \frac{110\,(y_1 - 2.7)}{0.8 \times 2.5} \times \frac{y_1 + 2.7}{2} \tag{3.31}$$

For $y_1 = 2.88$ m we find $m_{yf} = 27.6$; $m_{ys2} = -44.3$, which have a suitable ratio.

The moments for load distribution have their maxima at the centre of the loaded area, whereas the span moments in the strips have their maxima at x_1 and y_1. Adding together the moments for load distribution and the span moments in the strips will therefore give values slightly on the safe side. Performing this addition we get the following design moments:

$$m_{xf} = 24.2;\ m_{xs1} = -30.5;\ m_{yf} = 30.0;\ m_{ys2} = -44.3$$

This moment distribution is shown in Fig. 3.3.4. As pointed out in the previous example it is acceptable to distribute the support moments over larger widths.

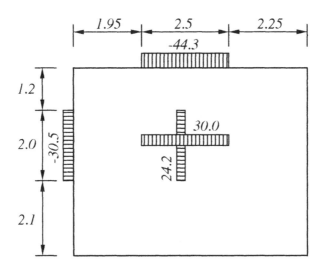

Fig. 3.3.4

3.3.3 Distributed and concentrated loads together

Example 3.9

The simply supported slab in Fig. 3.3.5 has a uniform load of 7 kN/m² and a point load of 40 kN in the position shown. The total uniform load is 7×6.1×4.7 = 200.7 kN. The point load is only 20% of the uniform load, which means that it can be treated in the simplified way recommended in 2.6.1 and 3.3.1. As it is difficult to estimate directly if it is to be taken only by element 2 or if it is distributed between elements 2 and 4 we start by assuming that the line of zero shear force passes through the point load but that the whole load is carried by element 2. We then get the following average moments from Eq. (2.5) for the distributed load and a simple moment equation for the point load:

$$m_{f2} = \frac{7 \times 1.8^2 (6.1 + 2 \times 1.9)}{6 \times 6.1} + \frac{40 \times 1.8}{6.1} = 17.94 \tag{3.32}$$

$$m_{f4} = \frac{7 \times 2.9^2 (6.1 + 2 \times 1.9)}{6 \times 6.1} = 15.92 \tag{3.33}$$

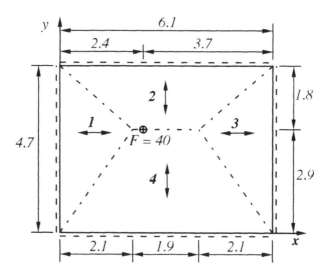

Fig. 3.3.5

From these values it can be seen that element 2 takes a little too much of the point load. It can easily be shown that the condition that the two values shall be identical is fulfilled if element 2 carries 37.39 kN and element 4 2.61 kN. The moment is $m_{yf} = 17.16$. If the above analysis had given $m_2 < m_4$ the line of zero shear force should have been moved to a position further down in the figure in order to make the moments equal.

A simplified way of calculating moments in this case is to treat the distributed load and the point load separately and add the maximum moments from these two cases, in spite of the fact that the maxima do not occur at the same sections. Such an addition is on the safe side. For symmetry reasons the line of zero shear force for the distributed load in this case is situated $4.7/2 = 2.35$ m from the edge. The average moment is

$$m_{yf} = \frac{7 \times 2.35^2 (6.1 + 2 \times 1.9)}{6 \times 6.1} + \frac{40 \times 1.8 \times 2.9}{4.7 \times 6.1} = 10.46 + 7.28 = 17.74 \qquad (3.34)$$

This value is about 3% higher than the value when the distributed load and the point load were treated together. This small difference is acceptable from the point of view of reinforcement economy. This simplified approach is satisfactory in many cases.

For the reinforcement in the x-direction Eq. (2.4) gives:

$$m_{xf} = \frac{7 \times 2.1^2}{6} = 5.15 \qquad (3.35)$$

The ratio between the moments in the x- and y-directions can be regarded as acceptable, but maybe a little low. If we wish to increase this ratio we may increase the height of the triangles and repeat the calculation.

We may calculate the distribution of design moments approximately as in Examples 3.1 and 3.2. However, in this case much of the bending moment is caused by the point load and from the point of view of performance in the service state it is better to concentrate more of the reinforcement closer to the point load.

At the same time another possibility can be demonstrated here, viz. zones of zero reinforcement along the supports. The slab has zero curvature along the supports and thus the reinforcement takes no stresses, at least not until at a late ultimate state when severe cracks appear and membrane action takes place. Therefore the reinforcement along the supports is of little use and it is better to use it where curvature exists. In the author's opinion such unreinforced zones should be used along supports in all slabs, but most building codes do not seem to accept this, which is why most examples in this book are shown without such zones.

In this example a width of 0.5 m has been chosen for the unreinforced zones and the reinforcement has been placed with a certain concentration around the point load. The proposed distribution of design moments is shown in Fig. 3.3.6. In the y-direction the smaller moments have been chosen to be one-third of the larger moments.

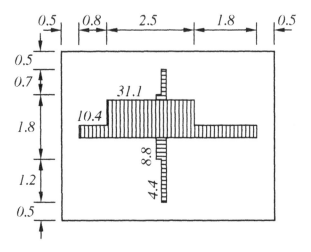

Fig. 3.3.6

CHAPTER 4

Rectangular slabs with one free edge

4.1 Introduction

4.1.1 General principles

As one edge is free (unsupported) the strips at right angles to that edge have no support at the free end. Instead they have to be supported internally, i.e. by a strip along the edge. Some solutions of this type are given in *Strip Method of Design*. The solutions shown below are somewhat different, as the design procedure adopted is similar to that used for slabs where all the sides are supported.

The support along the free edge is assumed to be a support band of the type described in Section 2.8, i.e. a band which in the analysis is treated as having zero width, but where the resulting reinforcement is distributed over a certain width along the edge. According to the recommendations in Section 2.8 the width over which the reinforcement is distributed depends on the importance of the support band to the safety of the slab. Typically this width can be chosen rather arbitrarily if the free edge is in the shorter direction of the slab, but it must satisfy some limiting rules if the free edge is in the longer direction of the slab. Some concentration of reinforcement along a free edge is generally recommended.

Fig. 4.1.1 a) shows a slab with the upper edge free. Lines of zero shear force are shown exactly the same as used for slabs with four sides supported. Element 2 has no real support. Instead it is assumed to be supported on a support band along the free edge. This support band is loaded by the load on element 2. The bending moment M_x in the strip is determined by the load and its distribution on element 2.

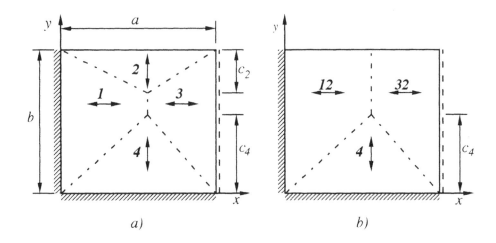

Fig. 4.1.1

For the total moment m_x in the slab we can add the moment in the strip to the moments in elements 1 and 3. Formally this can be done by analysing new elements 12 and 32, where the corresponding parts of element 2 are incorporated into elements 1 and 3. Thus the design of the slab can be based on the elements and lines of zero moments shown in Fig. 4.1.1 b).

A design based on Fig. 4.1.1 b) can only be used if $c_2 + c_4 \leq b$. The values of c_2 and c_4 depend on the moments m_{yf} and m_{ys} and on the type of load. For a uniform load it can be shown that this condition is fulfilled if

$$\frac{c_4}{b} \leq \frac{\sqrt{1 - m_{ys}/m_{yf}}}{1 + \sqrt{1 - m_{ys}/m_{yf}}} \tag{4.1}$$

In cases where this relation is valid it can be assumed that the width of the band of reinforcement is of minor importance for the safety of the slab. Thus in such a case no check of the part of the moment which is attributed to the band is needed, nor is any calculation of a suitable band of reinforcement or width of such a band needed. However, a certain concentration of reinforcement in the vicinity of the free edge is still recommended in such cases.

Where the above relationship is not fulfilled, elements 2 and 4 will no longer be triangles, but trapezoids, Fig. 4.1.2. The longer the line of zero shear force between elements 2 and 4 becomes, the more important is the role played by the support band in the safety of the slab. The moments in the support band must now be calculated and the width of the reinforcement band determined according to the principles discussed in Section 2.8.

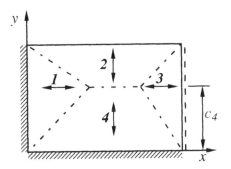

Fig. 4.1.2

4.1.2 Torsional moments. Corner reinforcement

The distribution of moments in slabs with an unsupported edge is more complicated than in slabs with all sides supported, particularly when the free edge is one of the longer edges. In such a case, torsional moments play an important role, whereas the bending moments corresponding to reinforcement at right angles to the free edge are of minor importance. In the strip method, solutions are used where the load is carried by bending moments only, without taking torsional moments into account. The resulting design is on the safe side regarding the ultimate limit state, but the distribution of reinforcement may be quite different from that given by an analysis based on the theory of elasticity.

According to the theory of elasticity large torsional moments occur where two simply supported edges meet. In these areas reinforcement may be necessary to prevent unacceptable cracking. Many codes specify particular corner reinforcement that should be provided.To limit cracks in the upper face of the slab the provision of top reinforcement in the slab corners parallell to the lines bisecting the corners is recommended. The amount of such reinforcement may be based on code rules or on values of torsional moments obtained from tables of moment values determined by the theory of elasticity.

The importance of the torsional moments increases as the ratio between the length of the free edge and the length of the perpendicular edges (*a/b* in Fig. 4.1.1) increases. The use of the form of the strip metod presented here is not recommended for simply supported slabs if this ratio is higher than about 2. An alternative approach for such slabs is demonstrated in *Strip Method of Design*.

73

4.2 Uniform loads

Example 4.1

The slab in Fig. 4.2.1 has one free edge, one simply supported edge and two fixed edges, and supports a uniform load of 9 kN/m^2. The shape is such that, with a relatively short free edge, the approach according to Fig. 4.1.1 b) may be used. The proposed lines of zero shear force are shown. For elements 12 and 32 we get from Eq. (2.5)

$$m_{xf} - m_{xs} = \frac{9 \times 2.7^2 (5.8 + 2 \times 3.2)}{6 \times 5.8} = 23.00 \tag{4.2}$$

$$m_{xf} = \frac{9 \times 1.6^2 (5.8 + 2 \times 3.2)}{6 \times 5.8} = 8.08 \tag{4.3}$$

and thus $m_{xs} = -14.92$. For element 4 we get from Eq. (2.4)

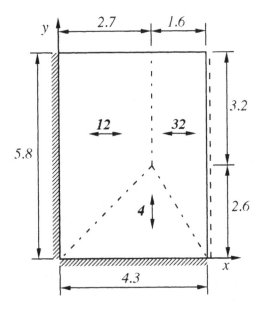

Fig. 4.2.1

$$m_{yf} - m_{ys} = \frac{9 \times 2.6^2}{6} = 10.14 \qquad (4.4)$$

For a fixed short edge it is suitable to choose a high ratio between the numerical values of the support and span moments. For example, in this case we may choose $m_{ys} = -7.5$ and $m_{yf} = 2.64$. The condition of Eq. (4.1) is well fulfilled.

These average moments can be distributed as shown in Fig. 4.2.2. The distribution of the reinforcement is arranged so that there is a strong band of reinforcement along the free edge.

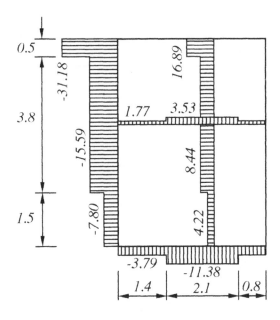

Fig. 4.2.2

Example 4.2

The slab shown in Fig. 4.2.3 carries a uniform load of 11 kN/m^2. It is of the type which cannot be designed by means of the simplified approach used in the previous example, as relation (4.1) cannot be fulfilled by adopting a suitable ratio between the moments in the x- and y-directions. Instead the slab has to be analysed as if it were simply supported on a support band located along its upper edge, after which an analysis must be made of the support band.

With the chosen pattern of lines of zero shear force and by applying Eqs. (2.4) and (2.5) as shown, we get the following average moments in the slab

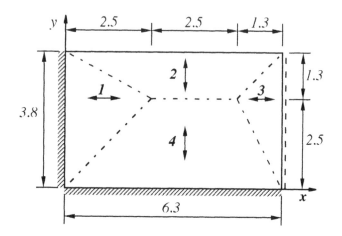

Fig. 4.2.3

$$m_{xf} = m_3 = \frac{11 \times 1.3^2}{6} = 3.10 \tag{4.5}$$

$$m_{xs} = 3.10 - \frac{11 \times 2.5^2}{6} = -8.36 \tag{4.6}$$

$$m_{yf} = \frac{11 \times 1.3^2(6.3 + 2 \times 2.5)}{6 \times 6.3} = 5.56 \tag{4.7}$$

$$m_{ys} = 5.56 - \frac{11 \times 2.5^2(6.3 + 2 \times 2.5)}{6 \times 6.3} = -14.99 \tag{4.8}$$

The load distribution on the support band corresponds to the shape of element 2. If we assume a section of zero shear force in the support band situated at a distance of 2.3 m from the right-hand end, we find the moments

$$M_f = 11 \times 1.3 \left[\frac{1.3}{2} \times \frac{2 \times 1.3}{3} + 1.0 \left(1.3 + \frac{1.0}{2} \right) \right] = 33.80 \tag{4.9}$$

$$M_s = 33.80 - 11 \times 1.3 \left[\frac{2.5}{2} \times \frac{2 \times 2.5}{3} + 1.5 \left(2.5 + \frac{1.5}{2} \right) \right] = -65.70 \tag{4.10}$$

If the recommendations in Section 2.8 are followed strictly, the span moment M_f will be concentrated on a width corresponding to half the average width of element 2, which corresponds to 0.45 m. The width has been chosen to 0.5 m in the proposed distribution of design moments shown in Fig. 4.2.4. It might even have been distributed over a larger width without any risk of inconvenience. In any case it is not possible to show by yield line theory that such a design is unsafe. The moments m_{xf} and m_{xs} have been distributed over the remaining part of the slab.

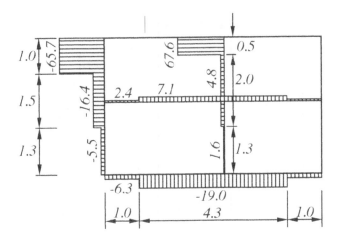

Fig. 4.2.4

4.3 Triangular loads

In practice, triangular load is usually a liquid pressure, an earth pressure or something similar. A rectangular slab with one free edge in such cases has the free edge as the upper edge. Only this case will be demonstrated here.

For the treatment of triangular loads with a combination of loading cases, see also Examples 3.5 and 3.6.

Example 4.3

The slab shown in Fig. 4.3.1 forms one side of a water tank with the highest water level 1.0 m below the upper edge. The water pressure is assumed to increase by 10 kN/m^2 per metre depth (a more correct value is 10.2). The distribution of the water pressure is shown in the figure. The shape of the slab is such that the simplified approach of Fig. 4.1.1 b) may be used. That this assumption is correct can afterwards be checked by studying a triangular element of type 2 in Fig. 2.1.1 a) and establishing that it does not reach element 4.

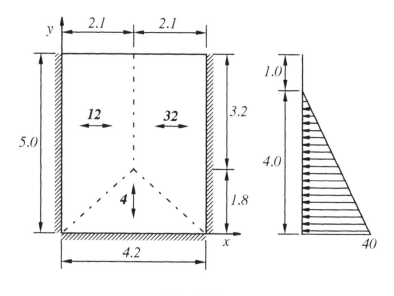

Fig. 4.3.1

The average moments in the x-direction can be calculated by means of Eq. (2.14) with $l = 4.0$, $l_1 = 0$, $l_2 = 2.2$, $l_3 = 1.8$. The average moment from this equation is valid for the depth 4.0 m. To get the average moment for the total depth 5.0 m, this value has to be multiplied by 4.0/5.0.

$$m_{xf} - m_{xs} = \frac{4.0}{5.0} \times \frac{40 \times 2.1^2}{24 \times 4.0^2} [6 \times 2.2^2 + 1.8 (4 \times 4.0 - 3 \times 1.8)] = 17.68 \qquad (4.11)$$

The average moments in the y-direction are calculated with Eq. (2.4) for a uniform load of 40 kN/m^2 minus a triangular load of 18 kN/m^2 at the top, calculated with Eq. (2.10).

78

$$m_{yf} - m_{ys} = \frac{40 \times 1.8^2}{6} - \frac{18 \times 1.8^2}{12} = 16.74 \qquad (4.12)$$

The following values of average moments are chosen to ensure suitable ratios between support and span moments: $m_{xf} = 5.68$, $m_{xs} = -12.00$, $m_{yf} = 4.74$, $m_{ys} = -12.00$. A distribution of design moments is proposed in Fig. 4.3.2. With respect to the load distribution, the m_x-moments might have been more concentrated downwards, but on the other hand there is also the general rule that recommends concentrating some reinforcement close to the free edge. As a compromise the design moment has been given a constant value over the major part of the depth.

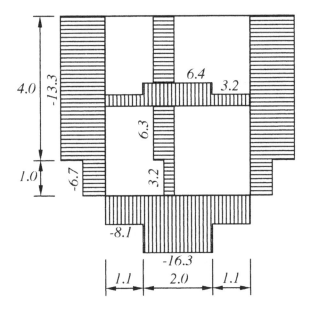

Fig. 4.3.2

Example 4.4

The slab in Fig. 4.3.3 forms one side of a water tank with the water surface 0.4 m below the upper edge. From that level the water pressure is assumed to increase by 10kN/m^2 per metre depth (a more correct value is 10.2). The shape of the slab is such that the simplified approach according to Fig. 4.1.1 b) cannot be applied. An assumed pattern of lines of zero shear force is shown in Fig. 4.3.3.

79

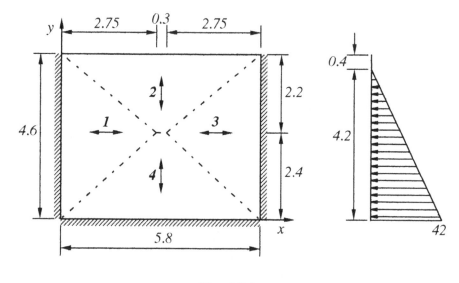

Fig. 4.3.3

In order to be able to use the standard formulas in Chapter 2 for calculating the moment in element 2 the load has to be divided into the following cases, which are added to form the actual load (cf. Example 3.6, Fig. 3.2.5):

1. A triangular load, zero at the free edge and 22 at the bottom.
2. A uniform load –4 over the whole depth.
3. A uniform load +4 on the uppermost 0.4 m.
4. A triangular load, zero at the free edge and –4 at 0.4 m below the free edge.

For the first and fourth cases Eq. (2.11) is applied, and for the second and third cases Eq. (2.5). For the third and fourth cases $l_1 = 4.80$ m, the width of element 2 at the water level.

$$m_{yf} = \frac{22 \times 2.2^2 (5.8 + 3 \times 0.3)}{12 \times 5.8} - \frac{4 \times 2.2^2 (5.8 + 2 \times 0.3)}{6 \times 5.8} +$$
$$+ \frac{4 \times 0.4^2 (5.8 + 2 \times 4.8)}{6 \times 5.8} - \frac{4 \times 0.4^2 (5.8 + 3 \times 4.8)}{12 \times 5.8} = 6.77 \qquad (4.13)$$

The moment in element 4 is calculated from the influence of a uniform load with intensity 42 according to Eq. (2.5) minus a triangular load with intensity 24 at the top of the element according to Eq. (2.11).

$$m_{yf} - m_{ys} = \frac{42 \times 2.4^2 (5.8 + 2 \times 0.3)}{6 \times 5.8} - \frac{24 \times 2.4^2 (5.8 + 3 \times 0.3)}{12 \times 5.8} = 31.18 \quad (4.14)$$

We get $m_{ys} = -24.41$.

The average moments in elements 1 and 3 can, with satisfactory accuracy, be calculated with a triangular load with maximum intensity 46 according to Eq. (2.13) minus a uniform load with intensity 4 according to Eq. (2.4). A small difference in the upper corner is disregarded.

$$m_{xf} - m_{xs} = \frac{46 \times 2.75^2 (4.6 + 2 \times 2.2)}{24 \times 4.6} - \frac{4 \times 2.75^2}{6} = 23.32 \quad (4.15)$$

We may then choose $m_{xs} = -16.0$; $m_{xf} = 7.82$.

The load on the support band along the free edge corresponds to the load on element 2. The load is looked upon as consisting of one part with a pyramidal shape and one part with a prismatic shape. Due to symmetry the point of zero shear force (maximum span moment) is situated in the centre.

$$M_f - M_s = \frac{18 \times 2.25 \times 1.8}{2 \times 3} (0.5 + \frac{3 \times 2.25}{4}) +$$
$$+ \frac{18 \times 0.15 \times 1.8}{2} (2.75 + \frac{0.15}{2}) = 33.44 \quad (4.16)$$

We can choose $M_f = 11.0$, $M_s = -22.44$.

To determine the design moments we have to add the moments from element 1 and from the support band. A distribution of design moments is proposed in Fig. 4.3.4. If the distributions from the different parts had been followed we should have had a concentrated band of reinforcement along the free edge, and also heavier reinforcement below the centre of the slab, as the load is increasing downwards. Halfway between the top and the centre the reinforcement would have been weaker. Such an uneven distribution of reinforcement does not correspond to our experience of the real behaviour in service. Therefore the design moment has been chosen with a uniform distribution over most of the depth, without any concentration.

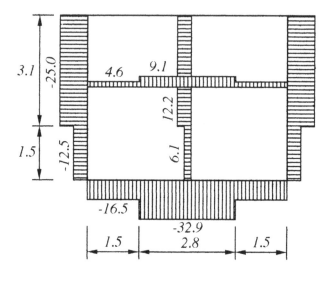

Fig. 4.3.4

4.4 Concentrated loads

4.4.1 Loads close to the free edge

A concentrated load acting close to a free edge has to be carried by reinforcement along the edge. If the load is acting very close to the edge it will cause a negative moment with a certain demand for top reinforcement at right angles to the edge. This reinforcement will in the first place be needed for the distribution of the load to a strip of a certain width in the ultimate limit state, but it may also be needed to limit top cracks. In principle, the strip method only takes the ultimate limit state into account, and this is the only case which will be demonstrated in the example below. In order to limit top cracks a minimum width of the assumed strip along the edge is recommended, e.g. 1/10 to 1/5 of the length of the free edge.

In practice the concentrated load always acts together with a distributed load, such as the load arising from the weight of the slab.

Example 4.5

The slab in Fig. 4.4.1 carries a point load of 30 kN at a position 0.1 m inside the edge. If we let 10 kN be carried to the right and 20 kN to the left support the moment in the edge strip is $M_f = 10 \times 2.5 = 25$ kNm, $M_s = -20 \times 3.0 + 25 = -35$ kNm. If we decide not to reinforce for a higher span moment than about 40 kNm/m for the point load (in addition to the moment for the other loads on the slab), we have to distribute the load over a strip width of about $25/40 = 0.625$ m. We can assume a width of 0.65 m.

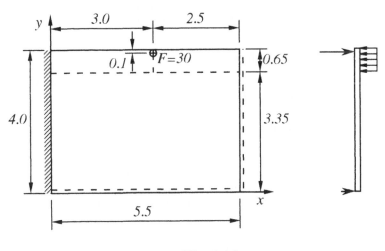

Fig. 4.4.1

A possible load distribution in the strip in the y-direction is shown in the figure to the right. In order to maintain equilibrium in the strip we need a downward force at the lower end of the strip, i.e. at the support. From simple statics this force is found to be $30 \times 0.225/3.675 = 1.8$ kN. In a rigorous analysis this force may be taken into account, but as it is small compared to the acting load, we will disregard its influence and simply take the design moment for the reinforcement at right angles to the edge as the load times the distance between the load and the centre of the strip, thus $M_y = -30 \times 0.225 = -6.8$ kNm. This analysis is quite accurate enough bearing in mind the approximate assumptions, which are conservative from the point of view of safety, as can be checked by means of yield line theory.

The result in this case is thus that the point load gives span reinforcement corresponding to $M_{xf} = 25$ kNm, distributed on a strip with a width of 0.65 m along the edge, and top reinforcement corresponding to $M_y = -6.8$ kNm at right angles to the edge at the load. The support moment $M_{xs} = -35$ kNm can be distributed over a greater width, as the support in itself has a load distributing effect. The moment M_y can be distributed on quite a small width, say 0.5-1.0 m in this case.

The use of the simplified approach which is demonstrated here is generally recommended for designing reinforcement to carry a concentrated load close to a free edge:

1. Calculate moments in the strip along the edge.
2. Determine a suitable width of the strip to give an acceptable concentration of span reinforcement.
3. Calculate the design moment for the top reinforcement at right angles to the edge as the load times the distance from the load to the centre of the strip. This reinforcement is concentrated on a rather small width near the load.

4.4.2 Loads not close to the free edge

In many cases it is simplest to assume that the load is carried by a strip which is parallel to the free edge. In this case the design is done exactly as in a slab with four sides supported, see Section 3.3.

When the distance from the load to the support opposite to the free edge is short, the load or part of it is best carried in the direction at right angles to the free edge. According to the general principles for slabs with a free edge, the strip at right angles to that edge is assumed to have a support at the edge, thus causing moments in a strip along the edge.

The example below demonstrates how the design is performed for a large concentrated load, which is the dominant load on the slab. Where the concentrated load is only a minor part of the total load on the slab a simplified approach may be used as was demonstrated in Example 3.9.

For the general treatment of concentrated loads see Sections 2.6 and 3.3.

Example 4.6

The slab in Fig. 4.4.2 carries a point load of 40 kN at a position which is closer to the support opposite to the free edge. This is a case where the load (or a great part of it) is best carried by a strip in the y-direction.

If the whole load is carried by a strip in the y-direction this strip is assumed to be supported at the free edge. We can assume, for example, that 6 kN is carried to the free edge. This gives a span moment $M_{yf} = 6 \times 3.0 = 18.0$ kNm. The support moment is then $M_{ys} = 18.0 - 34 \times 1.5 = -33.0$ kNm.

The edge strip has to carry the reaction 6 kN. If we assume that 2 kN is carried to the right support, the moments in the edge strip are: $M_{xf} = 2 \times 2.4 = 4.8$ kNm; $M_{xs} = 4.8 - 4 \times 3.6 = -9.6$ kNm.

If the whole load is instead carried by a strip in the x-direction we can then assume that 13 kN is carried to the right support, which gives $M_{xf} = 13 \times 2.4 = 31.2$ kNm in this strip and $M_{xs} = 31.2 - 27 \times 3.6 = -66.0$ kNm.

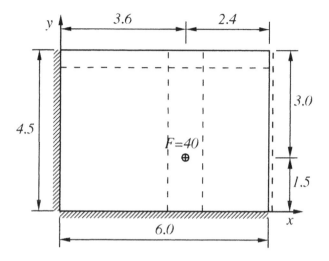

Fig. 4.4.2

If we compare the two solutions we find that the first solution gives a sum of span moments equal to $18.0 + 4.8 = 22.8$ compared to 31.2 in the second, and a numerical sum of support moments equal to $33.0 + 9.6 = 42.6$ in the first and 66.0 in the second solution. As the reinforcement area is proportional to the moment this comparison shows that the first solution is to be preferred from an economical point of view.

It is also possible to divide the load between the two solutions, for example, in order to get a better reinforcement distribution with respect to the behaviour under service conditions. If the concentrated load is dominant such a division is suitable, but if the distributed load on the slab is dominant such a division is unnecessary.

For a complete solution some reinforcement for load distribution across the strip should also be provided. If we choose a strip width of 1.0 m for the main strip and 0.5 m for the load distribution reinforcement, the design moment for this reinforcement is $m = 40 \times 1.0/(8 \times 0.5)$ $= 10.0$ kNm/m, Eq. (2.33). If we choose a width of 0.5 m for the strip along the free edge, we get the design moments according to the first solution as shown in Fig. 4.4.3. The support moments have been distributed over twice the widths of the strips. This is acceptable as the supports act as load distributors.

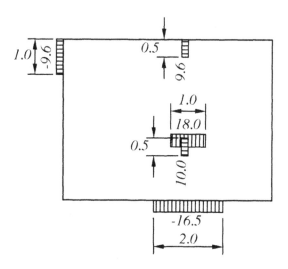

Fig. 4.4.3

CHAPTER 5

Rectangular slabs with two free edges

5.1 Two opposite free edges

When a rectangular slab has two opposite sides supported and the other two edges free, it acts as a series of one-way strips and causes no design problems, at least not as long as the loads are distributed.

Design for concentrated loads is made in the same way as for other rectangular slabs, see Section 3.3 for loads in the interior of the slab and Section 4.4 for loads close to a free edge.

5.2 Two adjacent free edges

5.2.1 General

Slabs with two adjacent free edges have a rather complicated static behaviour, particularly when the other two edges are freely supported. Because strips parallel to the edges do not have supports at both ends the load has to be carried to a great extent by torsional moments with respect to the edges.

There are at least two ways of coping with the problem. One possibility is to carry out an analysis which includes torsional moments and design reinforcement parallel to the edges to take the torsional moments. The slab can then be reinforced only with bars parallel to the edges, which is an advantage during construction. Solutions including torsional moments have been discussed in *Strip Method of Design*. These solutions are generally too complicated to be recommended for practical design. One such solution will, however, be given as an alternative below.

Another possible solution is to make use of a support band, which spans between the two corners where the free and supported edges meet. This method may lead to somewhat lengthy numerical calculations and checks. Its application is demonstrated in Chapter 7 on a non-rectangular slab. Here a similar but numerically simpler method will be applied in the first example, using corner-supported elements. The results of these two methods are similar, as the reinforcement is not parallel to the edges. This is a disadvantage from the point of view of construction, but it saves much reinforcement at the same time as it reduces deflections and cracks as the reinforcement is active approximately in the directions of the principal moments.

5.2.2 Simply supported edges, uniform loads

This is the most complicated case, as no load can be taken solely by bending moments parallel to the edges. It can even be argued that this is an unsatisfactory type of structure. It is probably not very common, at least not for structures other than small balconies, for example.

The centre of gravity of a uniform load on the slab is situated where the diagonals cross. It is therefore natural to assume that the slab is supported at point supports at the outer ends of the supported edges. If the slab is assumed to be supported along the supported edges some of the support reaction has to be negative in order to fulfil equilibrium conditions.

A possible solution is to assume that the slab is supported at the outer corners of the supported edges and to base the design on the use of corner-supported elements. Such a load-bearing system is illustrated in Fig. 5.2.1 with corner supports at B and D. The corner-supported elements are triangles formed by the edges and the diagonals, which are lines of zero shear force. In order to be able to use the rules for triangular corner-supported elements the reinforcement directions have to be parallel to the sides of the triangles, in this case parallel to the diagonals. It is impossible to maintain equilibrium on a line of zero shear force if the reinforcement is parallel to an edge of the slab, as the acceptable distributions of design moments in the two elements which meet at the line of zero shear force cannot then be made to coincide.

As the reinforcement directions in the general case are not at right angles, the rule in Section 2.5.4 has to be applied. This rule says that the c-values for both the reinforcement directions are equal to half the length of the diagonal.

The design of the reinforcement in direction B-D is based on Fig. 2.5.4 c) as the y-direction in this figure is parallel to the side which does not go to the supported corner. As there is no support moment in direction B-D, we find for that direction the average moment

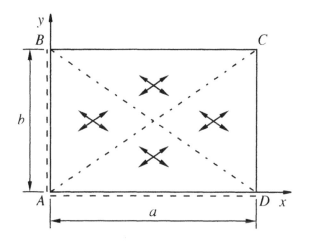

Fig. 5.2.1

$$m = \frac{q}{3} \left(\frac{\sqrt{a^2 + b^2}}{2} \right)^2 = \frac{q(a^2 + b^2)}{12} \tag{5.1}$$

For the reinforcement in direction *A-C* there is no span moment. The design is based on Fig. 2.5.4 b). We find for direction *A-C* the average moment

$$m = -\frac{q(a^2 + b^2)}{24} \tag{5.2}$$

For the distribution of design moments we have to follow the rules given in Section 2.5.3.

For direction *B-D* Eq. (2.32) is valid. In this direction we have to take all the design moment only in the "column strip" as $\alpha = 0$. It may be suitable to distribute the reinforcement as evenly as possible, i. e. with $\beta = 1/3$. This means that the design moment is zero in the exterior part of the element and 3 times the average moment in the central part ("column strip"). It is not acceptable to leave the exterior parts without reinforcement, so some reinforcement must also be provided in these parts. In Fig. 5.2.2 a design moment is recommended equal to half the average moment according to Eq. (5.1) within these parts. This choice can of course be discussed.

For direction A-C Eq. (2.31) is valid. Here it is acceptable to use $\alpha = 1.0$ which means an evenly distributed design moment equal to the average moment, which is a suitable choice in this case.

Using these values we get the design moments according to Fig. 5.2.2. It must be noted that the rules in Section 2.4.3 are based on the assumption that all reinforcing bars have a full length through the element and are well anchored at the edges. It is also recommended to have at least one bottom bar and one top bar along each free edge.

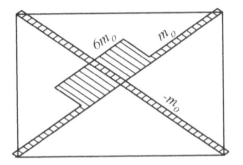

$$m_o = q(a^2 + b^2)/24$$

Fig. 5.2.2

As the reinforcement based on this design is complicated, involving different lengths for each bar, a design with bars parallel to the edges may be preferred, even if it leads to a greater amount of reinforcement. Applying a solution including torsional moments of the type discussed in *Strip Method of Design*, Section 4.3, it can be shown that a safe design would be to reinforce the whole slab parallel to the edges for the positive moments

$$m_x = m_y = \frac{qab}{4} \tag{5.3}$$

and the negative moments

$$m_x = m_y = -\frac{qab}{2\left(2 + \dfrac{b}{a}\right)} \quad \text{for } a \geq b \tag{5.4}$$

As the reinforcement is intended to take torsional moments it has to be well anchored at the edges.

In order to fulfil equilibrium conditions with this solution the inner corner *A* in Fig. 5.2.1 has to be anchored by a force *R* equal to

$$R = \frac{qab}{2 + \dfrac{b}{a}} \qquad \text{for } a \geq b \tag{5.5}$$

The total amount of design reinforcement is more than doubled with this design, but in spite of this it may often be recommended due to the simplicity of construction. Due regard has also to be taken of the fact that the first solution may be regarded as unacceptable without some secondary reinforcement.

5.2.3 One fixed edge, uniform loads

If a support is fixed, a strip at right angles to that support can act as a cantilever, carrying load by means of pure bending moments. In principle all the load can be carried in this way if at least one of the supports can take negative moments. If only one of the supports is fixed such a solution is uneconomical and it also gives a design which is not suitable for the behaviour of the slab under service conditions.

If all the load is carried by cantilever action there will be only top reinforcement in the slab. This is not acceptable, as it may result in wide bottom cracks. In the service state moments will give rise to tension in the bottom face of the slab. A simple way of taking care of this problem is to assume that one part of the load is carried without cantilever action, i.e. as if the slab were simply supported. The moments caused by that part of the load are then calculated by means of equations (5.3) and (5.4). The amount of the load carried in this way is chosen to ensure that a suitable minimum amount of bottom reinforcement is provided.

Example 5.1

The slab in Fig. 5.2.3 has one long fixed support and one short free support. The load is uniform 9 kN/m². In order to get some bottom reinforcement in both directions it is assumed that 20% of the load gives moments according to Eqs (5.3) and (5.4). This ratio should be chosen so that the amount of bottom reinforcement is sufficient to prevent wide bottom cracks and to take at least some of the torsional moments in the service state.

The part of the load to be used in Eqs (5.3) and (5.4) is thus $0.2 \times 9 = 1.8$ kN/m². This gives design moments, which are active in the whole slab:

$$m_x = m_y = \frac{1.8 \times 6.0 \times 4.0}{4} = 10.80 \tag{5.6}$$

$$m_x = m_y = -\frac{1.8 \times 6.0 \times 4.0}{2\left(2 + \dfrac{4.0}{6.0}\right)} = -8.10 \tag{5.7}$$

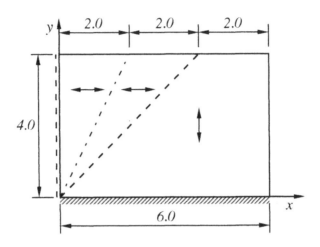

Fig. 5.2.3

The rest of the load, 7.2 kN/m^2, is carried by one-way strips as shown in Fig. 5.2.3. The strip in the x-direction is supported at its right-hand end by the cantilever in the y-direction along the dashed line, which is the limit between the x- and y-strips. The line of zero shear force is halfway between the dashed line and the left-hand support.

As the cantilever in the y-direction has to act as a support for the x-strip, it has to carry all the load to the right of the line of zero shear force. The average support moment is calculated by means of Eq. (2.5)

$$m_{ys} = -\frac{7.2 \times 4.0^2 (6.0 + 2 \times 4.0)}{6 \times 6.0} = -44.80 \qquad (5.8)$$

The average span moment in the x-strip is calculated by means of Eq. (2.4)

$$m_{xf} = \frac{7.2 \times 2.0^2}{6} = 4.80 \qquad (5.9)$$

The total moments are the sum of the moments from the two cases, thus $m_{xs} = -8.10$ (not a real support moment, but a negative moment), $m_{xf} = 4.80 + 10.80 = 15.60$; $m_{ys} = -44.80 - 8.10 = -52.90$; $m_{yf} = 10.80$. The design moments according to case 1 should be evenly distributed, whereas the design moments according to case 2 should be distributed with respect to the shape of the elements, i.e. with more reinforcement along the free edges. Distributions of design moments for bottom reinforcement in the x-direction and top reinforcement in the y-direction are proposed in Fig. 5.2.4. The design moments for top reinforcement in the x-direction and for the bottom reinforcement in the y-direction are constant with the values above. The reinforcement corresponding to case 1 should cover the whole slab area and be well anchored, whereas the reinforcement corresponding to case 2 may be curtailed according to normal detailing rules.

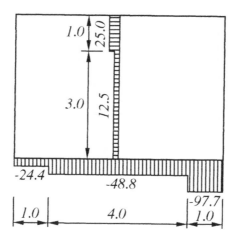

Fig. 5.2.4

Example 5.2

The slab in Fig. 5.2.5 has the same proportions and the same load, 9.0 kN/m², as the slab in the previous example, but it is the short edge that is fixed and the long edge which is simply supported. This leads to behaviour expected to be closer to that of a slab where both edges are simply supported. It is then natural to carry more of the load as if the slab were simply supported. Whereas in the previous example we chose to carry 20% of the load in this way, here we choose to take half the load in this way. Thus 4.5 kN/m² is assumed to give moments according to Eqs (5.3) and (5.4):

$$m_x = m_y = \frac{4.5 \times 6.0 \times 4.0}{4} = 27.00 \qquad (5.10)$$

$$m_x = m_y = -\frac{4.5 \times 6.0 \times 4.0}{2\,(2 + \frac{4.0}{6.0})} = -20.25 \qquad (5.11)$$

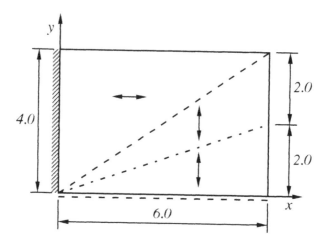

Fig. 5.2.5

The remaining part of the load, 4.5 kN/m², is carried by the strip system according to Fig. 5.2.5, which is similar to the system in Fig. 5.2.3 with the exception that the dividing line between the primarily load-bearing directions is drawn to the outer corner. If it had been drawn to a position more to the left, the strips in the y-direction would not have had a support on their whole width. If it had been drawn to a lower position at the right-hand edge the moment in the strip would have increased too much.

In the same way as in the previous example the moments are calculated by means of Eqs. (2.5) and (2.4) respectively:

$$m_{xs} = -\frac{4.5 \times 6.0^2 (4.0 + 2 \times 2.0)}{6 \times 4.0} = -54.00 \qquad (5.12)$$

$$m_{yf} = \frac{4.5 \times 2.0^2}{6} = 3.00 \qquad (5.13)$$

Adding these moments to those of the first case we get $m_{xs} = -74.25$ and $m_{yf} = 30.00$. Distributions of the corresponding design moments are proposed in Fig. 5.2.6. In addition to these moments the steel for evenly distributed reinforcement should be arranged for design moments m_{xf} and m_{ys} according to Eqs (5.10) and (5.11) above.

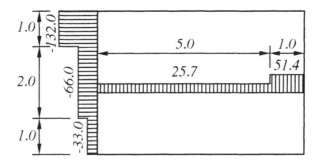

Fig. 5.2.6

5.2.4 Two fixed edges, uniform loads

With both supported edges fixed it is simplest and most economical to carry the whole load on cantilevering strips from those supports. With this solution the slab only requires top reinforcement, but this may not ensure satisfactory behaviour under service conditions. In order to provide for some bottom reinforcement in the design some of the load can instead be carried as if the supports were freely supported, i.e. applying Eqs (5.3) and (5.4) for that part of the load.

The dividing line between the cantilevers in the two directions is best drawn approximately in the direction of the bisector between the supported edges.

Example 5.3

The slab in Fig. 5.2.7 has the same proportions and the same load, 9 kN/m^2, as the slabs in the previous two examples, but both supported edges are fixed.

In order to get some bottom reinforcement we take 20% of the load, 1.8 kN/m^2, as if the slab were simply supported, applying Eqs (5.3) and (5.4). We can take the values directly from Example 5.3, where the same assumption was used:

$m_x = m_y = 10.80$ and $m_x = m_y = -8.10$.

The rest of the load, 7.2 kN/m^2, is carried by the cantilevers. From Eqs (2.4) and (2.5) we derive the average support moments

$$m_{xs} = -\frac{7.2 \times 4.0^2}{6} = -19.20 \qquad (5.14)$$

$$m_{ys} = -\frac{7.2 \times 4.0^2 (6.0 + 2 \times 2.0)}{6 \times 6.0} = -32.00 \qquad (5.15)$$

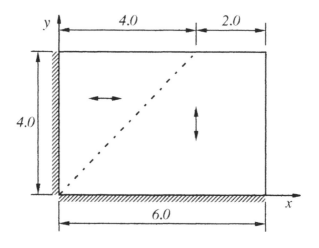

Fig. 5.2.7

Adding the negative values above gives $m_{xs} = -27.30$; $m_{ys} = -40.10$. Distributions of the negative design moments are proposed in Fig. 5.2.8. In addition to these negative moments the whole slab should be reinforced in both directions for positive moments with intensity 10.80.

5.2.5 Non-uniform loads

The same principles as those above can be applied to other types of distributed loads. The principle of Fig. 5.2.1 is complicated to apply, partly because the shapes of the corner-supported elements have to be modified in order to fulfil equilibrium conditions, and partly because the rules for acceptable moment distributions are not known in detail. If the principle according to Fig. 5.2.1 is applied this has to be done with care, using design moments which are estimated to be well on the safe side.

It is always possible to use the approach with a support band, which is demonstrated in Section 7.2.4.

For slabs with fixed supports the principles demonstrated in Examples 5.1-3 can be applied to any type of distributed load. The part carried by applying Eqs. (5.3) and (5.4) is always chosen to be uniform.

Concentrated loads on slabs with fixed supports can be treated with the methods demonstrated in Sections 3.3 and 4.4.

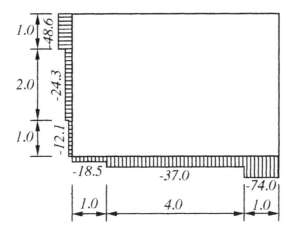

Fig. 5.2.8

A concentrated load F kN on a slab with two simply supported edges can be assumed to give rise to design moments equal to ±F/2 kNm/m within a rectangle with sides along the supported edges and a corner at the centre of the load, see Fig. 5.2.9.

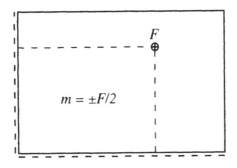

Fig. 5.2.9

As non-uniform loads on slabs with two adjacent free edges are rarely encountered, no numerical example is given, but it is hoped that the advice given above will be sufficient to enable a safe design to be carried out in any such cases.

97

CHAPTER 6

Triangular slabs

6.1 General

6.1.1 Reinforcement directions

At fixed or continuous supports the most efficient use of reinforcement is with the bars at right angles to the support. Span reinforcement is suitably arranged as a rectangular mesh with one of the mesh directions parallel to one of the slab edges. If the slab has one free edge it is natural to place the span reinforcement parallel to that edge. In other slabs it is often suitable to choose one of the span reinforcement directions to be parallel to the shortest side.

6.1.2 Calculation of average moments in whole elements

Whereas in a rectangular slab all reinforcement is normally arranged in only two orthogonal directions, parallel to and at right angles to the edges, this is not possible in triangular slabs. This poses particular problems. Reinforcement oriented in different directions has to cooperate. Some reinforcement forms a skew angle to the edge where the corresponding element has its support.

The basic method of analysis is the same as for rectangular slabs. The slab is divided into elements by means of assumed lines of zero shear force so that each element has its support along one side of the triangle. For each element the moment equilibrium is considered with respect to the support. The difference from rectangular slabs is that there is not one well defined load-bearing direction in each element, as the load has to be carried by means of reinforcement running in different directions. For example, these may be one direction for the top reinforcement and one or two directions for the bottom reinforcement. There are two different approaches for determining the design moments under these circumstances, see Section 2.3.6.

One approach is to introduce a line of zero moment, with positive moments on one side and negative moments on the other. Span reinforcement is active within the part with positive moment and determines the load-bearing direction within that part, whereas the support reinforcement determines the load-bearing direction within the part with negative moments. Strips with span reinforcement are treated as supported on cantilevers with support reinforcement. With this approach the moment distribution can be determined in a rigorous way. The numerical calculations may become complicated, particularly when the same support reinforcement functions together with the span reinforcement in two directions.

The other approach is to write a complete equilibrium equation with respect to the support for each element, taking into account the moment vectors along all the boundaries of the element. The numerical calculations are generally simpler with this approach. The disadvantages are that the result does not give a clear overview of the details of how the load is carried, or of the theoretical moment distributions which would be used as a basis for determining the distribution and curtailment of reinforcement.

6.1.3 Distribution of reinforcement

Where a reinforcing bar cuts over a corner, the length of the bar within the slab is very short near the corner. A bar with such a short length cannot be expected to be of any use. It is therefore proposed that, for slabs of normal thickness (about 0.15 to 0.2 m) bars shorter than about 0.5 m should not be used within the slab. The design moments within such parts of a slab are thus reduced to zero.

Elements in triangular slabs have triangular shapes. The theoretical moment distribution is more uneven than in elements of trapezoidal shape, which carry most of the load in rectangular slabs, see Figs. 2.3.3 and 2.3.4. In the proposed distribution of design moments in the examples the ratio between the major and minor design moments has been chosen to be 3, whereas for rectangular slabs this ratio was often chosen to be 2.

At corners there may be a need for top reinforcement running approximately in the direction of the bisector of the corner angle in order to avoid top cracks. Such reinforcement cannot be derived from the solutions given by the strip method, but has to be designed by some other means, based on the theory of elasticity.

6.2 Uniform loads

6.2.1 All sides simply supported

Example 6.1

The simply supported triangular slab in Fig. 6.2.1 has a uniform load of 8 kN/m². The assumed lines of zero shear force are shown.

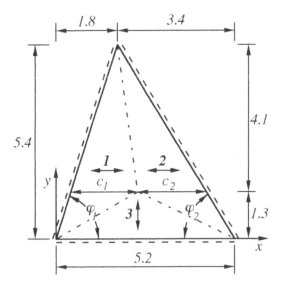

Fig. 6.2.1

It has been shown in *Strip Method of Design* that a suitable height of element 3 is 0.2 to 0.25 of the length of the short side. In this case it has been chosen to be $0.25 \times 5.2 = 1.3$ m.

First an approximate solution will be shown, where we use the approximation demonstrated in Fig. 2.2.4 for element 3 (which means that the dash-dot lines bordering element 3 are not real lines of zero shear force, though they are treated as such) and also disregard the influence of m_{yf} on the equilibrium of elements 1 and 2. This means that both elements 1 and 2 will cause the same span moment m_{xf}, calculated from Eq. (2.4) with c being the horizontal distance from the point of intersection of the lines of zero shear force to the edge, cf. Fig. 2.3.15. In order to get the same values of m_{xf} the distances c_1 and c_2 have to be equal. We thus find

$$c_1 = c_2 = \frac{4.1}{5.4} \times \frac{5.2}{2} = 1.974 \tag{6.1}$$

$$m_{xf} = \frac{8 \times 1.974^2}{6} = 5.20 \tag{6.2}$$

$$m_{yf} = \frac{8 \times 1.3^2}{6} = 2.25 \tag{6.3}$$

This is evidently a very simple way of determining the average moments. The approximation is on the safe side. It may be noted that the horizontal position of the peak of the triangle does not influence the result.

For a more rigorous solution we have to take into account the influence of m_{yf} on the equilibrium of elements 1 and 2, and study the equilibrium with respect to the supports. Fig. 6.2.2 shows element 1 with the acting total moments at the boundaries. Applying Eq. (2.18) and Eq. (2.4) we get

$$m_{xf} = \frac{8c_1^2}{6} - m_{yf}\cot^2\varphi_1 \tag{6.4}$$

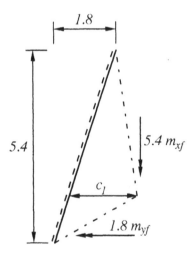

Fig. 6.2.2

The last term in this equation is the reduction in m_{xf} due to m_{yf}. In the same way we find for element 2

$$m_{xf} = \frac{8c_2^2}{6} - m_{yf}\cot^2\varphi_2 \tag{6.5}$$

We also have the relation

$$c_1 + c_2 = \frac{4.1}{5.4} \times 5.2 = 3.948 \tag{6.6}$$

With $m_{yf} = 2.25$ we can solve these equations and find $c_1 = 1.913$, $c_2 = 2.035$ and $m_{xf} = 4.63$, which is a lower value than the approximate value of 5.20 above. An estimation of the relative amounts of reinforcement with the two solutions can be made by comparing the sum of moments in the two directions, which gives the ratio $(5.20 + 2.25)/(4.63 + 2.25) = 1.08$. The simple approximate solution thus requires about 8% more reinforcement because it is more conservative.

In Fig. 6.2.3 a distribution of design moments has been proposed, based on the more accurate and economical solution, and taking into account the general rules for reinforcement distribution in Section 6.1.3. The theoretical distribution of moments m_{xf} is also shown.

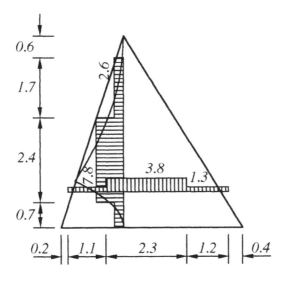

Fig. 6.2.3

Lastly it should be mentioned that a very simple solution for a simply supported triangular slab with a uniform load is possible if the same average moments are used in two orthogonal directions. This average moment is

$$m_{xf} = m_{yf} = \frac{qr_i^2}{6} \tag{6.7}$$

where r_i is the radius of the inscribed circle. The point of intersection of the lines of zero shear force is at the centre of the circle. For the triangle in the example $r_i = 1.626$, which gives $m_{xf} = m_{yf} = 3.52$. The total amount of reinforcement with this solution is a few percent higher than with the most accurate solution above, but lower than with the simple approximate solution.

6.2.2 One free edge

Example 6.2

The slab in Fig. 6.2.4 has two orthogonal fixed supports and one free edge. It is acted upon by a uniform load of 7 kN/m². The top reinforcement is at right angles to the supports and the bottom reinforcement is parallel to the free edge. Thus there is designed bottom reinforcement only in the direction parallel to the free edge. The elements are denoted 1 and 2, the average support moments m_{s1} and m_{s2}, and the span moment m_{xf}.

A coordinate system with the x-axis in the direction of the span bottom reinforcement, parallel to the free edge, is introduced in order to be able to use Eq. (2.18). Applying that equation and Eq. (2.4) and using the notation in the figure we have the equilibrium equations

$$m_{xf} - \frac{m_{s1}}{\sin^2 (\pi/2 + \varphi)} = \frac{7c_1^2}{6} \tag{6.8}$$

$$m_{xf} - \frac{m_{s2}}{\sin^2 \varphi} = \frac{7c_2^2}{6} \tag{6.9}$$

We also have the relation

$$c_1 + c_2 = \sqrt{4.0^2 + 6.0^2} = 7.21 \tag{6.10}$$

Assuming values of m_{xf} and c_1, the corresponding support moments can be calculated. The choice is made such that suitable ratios between the moments result. With $m_{xf} = 2.8$ and

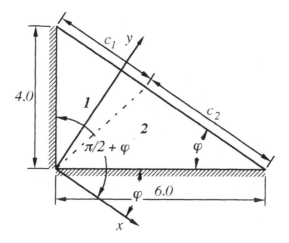

Fig. 6.2.4

$c_1 = 3.10$ the support moments are $m_{s1} = -5.82$, $m_{s2} = -5.20$. These moments are used for the design in Fig. 6.2.6.

In order to demonstrate the use of cooperating strips and how a theoretical moment distribution can be determined, an analysis will also be made with this method, using the same average moments. For this purpose lines of zero shear force and zero moments are shown in Fig. 6.2.5. Elements 1 and 2 are divided into span elements 1f and 2f and support elements 1s and 2s. Each of these elements now has only one load-bearing direction.

With $m_{xf} = 2.8$ and $q = 7$ the c-value of elements 1f and 2f are, cf. Eq. (2.4):

$$c_{1f} = c_{2f} = \sqrt{\frac{6 \times 2.8}{7}} = 1.55 \tag{6.11}$$

The distances shown in Fig. 6.2.5 are calculated by means of simple geometry.

The shear forces at the lines of zero moment are proportional to the distance from the inner corner of the slab. We can study a thin span strip along the edge. At the ends of this strip the shear force is

$$Q_{1f} = Q_{2f} = 7 \times 1.55 = 10.85 \text{ kN/m} \tag{6.12}$$

The corresponding shear forces at the ends of the support strips are calculated by means of Eq. (2.16). Inserting the relevant sin-values found from ordinary trigonometrical relations, we find:

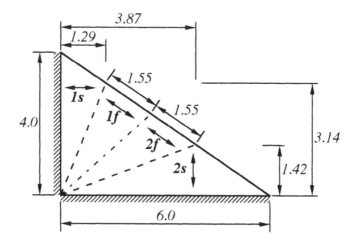

Fig. 6.2.5

$$Q_{1s} = \frac{0.9804}{0.9250} \times 10.85 = 11.50 \qquad (6.13)$$

$$Q_{2s} = \frac{0.8074}{0.9388} \times 10.85 = 9.33 \qquad (6.14)$$

The average moments caused by these shear forces on the relevant parts of the supports are according to Eq. (2.15)

$$\Delta m_{s1} = -\frac{11.50 \times 1.29}{3} = -4.95 \qquad (6.15)$$

$$\Delta m_{s2} = -\frac{9.33 \times 1.42}{3} = -4.42 \qquad (6.16)$$

The average moments caused by the direct load on the support elements give

$$m_{s1} = -\frac{7 \times 1.29^2}{6} = -1.94 \qquad (6.17)$$

$$m_{s2} = -\frac{7 \times 1.42^2}{6} = -2.35 \qquad (6.18)$$

The sums of average moments along the supports are

$$m_{s1} = -\frac{3.14}{4.0} \times 4.95 - 1.94 = -5.82 \tag{6.19}$$

$$m_{s2} = -\frac{3.87}{6.0} \times 4.42 - 2.35 = -5.20 \tag{6.20}$$

These values are identical with the values determined for elements 1 and 2 above.

It may be noted from the analysis that the influence of the shear force along the line of zero moment dominates in calculating the support moments. This has to be taken into account in distributing the design moments.

A distribution of design moments has been proposed in Fig. 6.2.6. In order to illustrate the influence of the different parts of the support moment, a curve has been drawn for the theoretical distribution of m_{s2} according to the model with cooperating strips. It can be seen that this model gives a steep jump in the moment intensity. This jump does not have much to do with the real behaviour of the slab, and there is no point in trying to reinforce exactly according to this distribution. The curve does, however, indicate which part of the edge most of the reinforcement should be placed in.

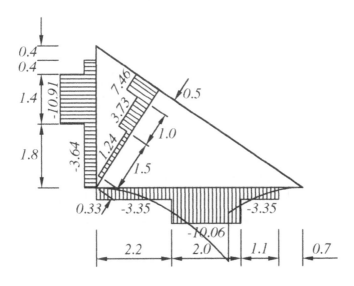

Fig. 6.2.6

Regarding the span reinforcement it is always advisable to have a certain concentration of reinforcement along a free edge.

In this case the slab has design bottom reinforcement in only one direction. Some secondary reinforcement might be introduced at right angles to the design reinforcement according to relevant code rules. In the author's opinion it is questionable whether such reinforcement is of any use.

6.2.3 Fixed and simply supported edges

Example 6.3

The slab in Fig. 6.2.7 has two fixed edges and one simply supported edge. The load is 9 kN/m^2. The lines of zero shear force are shown with the c-values of the three elements inserted in the x- and y-directions respectively. The bottom reinforcement is arranged in the x- and y-directions. From Eqs (2.18) and (2.4) we get the equilibrium equations

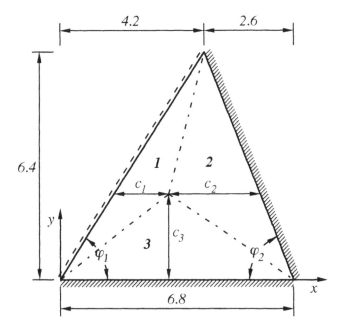

Fig. 6.2.7

$$m_{xf} + m_{yf}\cot^2\varphi_1 = \frac{9c_1^2}{6} \tag{6.21}$$

$$m_{xf} + m_{yf}\cot^2\varphi_2 - \frac{m_{s2}}{\sin^2\varphi_2} = \frac{9c_2^2}{6} \tag{6.22}$$

$$m_{yf} - m_{s3} = \frac{9c_3^2}{6} \tag{6.23}$$

We also have the geometrical relation

$$c_1 + c_2 = \frac{6.4 - c_3}{6.4} \times 6.8 \tag{6.24}$$

After two c-values and the ratio m_{s3}/m_{yf} have been chosen all the corresponding moments can be calculated. The choice is repeated until the ratios between the moments are estimated to be acceptable. A solution with $c_1 = 1.65$, $c_3 = 2.4$ gives $c_2 = 2.60$, $m_{xf} = 2.88$, $m_{yf} = 2.80$; $m_{s2} = -5.83$, $m_{s3} = -5.84$.

In this case it is complicated to make a design based on the cooperation of strips. The possibility exists of using the approximate solution where element 3 is treated according to the approximation described in Section 2.2, Fig. 2.2.4, but this gives a solution which is unnecessarily conservative. Based on the shape of the elements it is however possible to estimate a suitable moment distribution.

A moment distribution based on the values given above and taking into account the points of view in Section 6.1.3 is proposed in Fig. 6.2.8.

6.3 Triangular loads

Example 6.4

Fig. 6.3.1 shows a wall in the shape of a triangular slab with one fixed edge, one simply supported edge and one free edge. The wall is acted upon by an earth pressure, which can be assumed to be zero at the free edge and to increase by 8 kN/m^2 per vertical metre. This means that the load on the slab has a linear variation in both the vertical and the horizontal directions, with a maximum intensity of $3.5\times8 = 28$ kN/m^2 at the inner corner.

The wall is reinforced for a positive moment with bars parallel to the free edge and for a negative moment at the fixed support with bars at right angles to that support. The assumed line of zero shear force is shown in the figure.

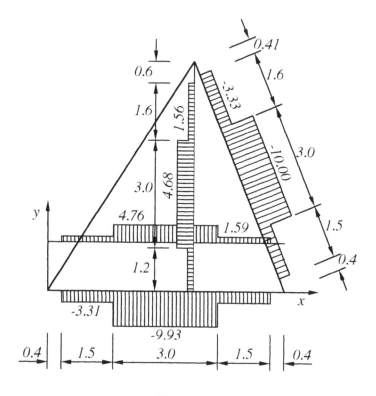

Fig. 6.2.8

The analysis of this slab follows exactly the same lines as Example 6.2 above. The only difference is that the loading cases for both elements correspond to Fig. 2.3.11 with $l_1 = 0$. We thus use Eq. (2.18) together with 2.13, which gives the equations

$$m_{xf} - \frac{m_{s1}}{\sin^2 (\pi/2 + \varphi)} = \frac{28c_1^2}{24} \qquad (6.25)$$

$$m_{xf} = \frac{28c_2^2}{24} \qquad (6.26)$$

We also have the geometrical relation

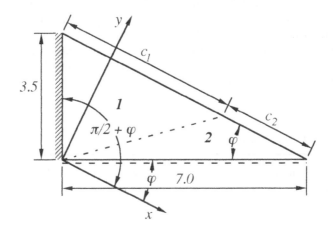

Fig. 6.3.1

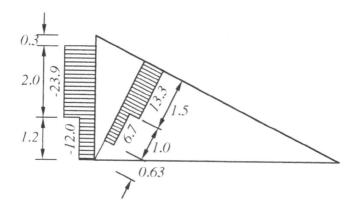

Fig. 6.3.2

$$c_1 + c_2 = \sqrt{3.5^2 + 7.0^2} = 7.83 \qquad (6.27)$$

and

$$\sin^2(\pi/2+\varphi) = 0.8. \qquad (6.28)$$

111

Choosing $c_1 = 5.13$, $c_2 = 2.70$ we find $m_f = 8.51$, $m_{sl} = -17.76$. A distribution of design moments is proposed in Fig. 6.3.2. As the load is zero along the free edge there is no reason to use a strong band of reinforcement along the edge, but on the other hand the reinforcement should not be reduced along the edge. Beside the reinforcement which is designed for these moments some minimum transverse bottom reinforcement may be required by codes, as well as extra top reinforcement at the corner.

6.4 Concentrated loads

The design for a concentrated load on a triangular slab is made with the same methods which are used for rectangular slabs, see Sections 3.3 and 4.4.

CHAPTER 7

Slabs with non-orthogonal edges

7.1 General

This chapter treats slabs which are neither rectangular nor triangular and where the corners between edges have angles smaller than 180°. Corners with an angle greater than 180° are called reentrant corners. Slabs with reentrant corners are treated in Chapter 11.

The general rules for analysis are the same as for rectangular and triangular slabs. Examples of analyses are only given for uniform loads, as such loads are the most common. In cases where other types of load are acting on a slab modifications to take these into account can be based on the rules and examples given for rectangular and triangular slabs.

Support reinforcement is most efficient if it is arranged at right angles to the direction of the support. This has been assumed to apply generally.

Span reinforcement is normally assumed to be arranged in two orthogonal directions, parallel to the chosen coordinate axes. In some cases one part of the span reinforcement is arranged in another direction. This applies particularly with free edges, where some span reinforcement is always placed along the free edge.

As in the earlier chapters the analysis is based on the assumption of a pattern of lines of zero shear force. These lines should, in principle, be so arranged that the amount of reinforcement needed to carry the load is as small as possible. This generally means that if the support conditions are the same (both freely supported or both fixed) the distances from a line of zero shear force to the two nearest supports should be chosen to be approximately equal. Where the support conditions are different the distance to the fixed support should be 1.5 to 2 times the distance to the simple support. In a direction with longer span reinforcement the distance shall be chosen to be shorter compared to a direction with a shorter span reinforcement. The final choice is based on an estimate of the suitability of the ratio between different moments.

By means of the lines of zero shear force the slab is divided into elements. Each element is bounded by a number of lines of zero shear force and one straight support. The equilibrium equation for each element is established, e.g. by means of Eq. (2.18).

By means of the equilibrium equations and geometrical conditions the average moments can be calculated for an assumed pattern of lines of zero shear force. This pattern will often have to be modified a number of times before the ratio between the different moments is considered to be satisfactory. *In the examples below only the final pattern, which is the result of such modifications, will be shown.*

During the analyses the moments are generally treated as if they were uniformly distributed in the lateral direction. The final distribution of design moments is however chosen with respect to an estimated more correct theoretical distribution, with greater moments in the central parts. This procedure is satisfactory from the point of view of safety.

7.2 Four straight edges

7.2.1 All edges supported

At each of the corners a line of zero shear force will start. There will usually also be one such line which does not start at a corner. The direction of this line is best chosen approximately in the direction of the bisector to the supports of the elements which are separated by the line. In this way the pattern of lines of zero shear force is at the same time a yield line pattern, which shows that the design is reasonably economical.

Example 7.1

The slab in Fig. 7.2.1 has two fixed and two freely supported edges. The top reinforcement is at right angles to the supports and the bottom reinforcement is parallel to the x- and y-axes. The load is 9 kN/m^2.

As the slab has a rather irregular shape the x- and y-coordinates have been given for the corners and for the crossing points of the lines of zero shear force. The pattern of the lines of zero shear force follows the general rules given above. It is the result of a trial and error process until the ratios between the moments are estimated to be acceptable.

With this pattern of lines of zero shear force we get the following c-values:
$c_1 = 1.55$, $c_{21} = 1.70$, $c_{22} = 1.80$, $c_3 = 3.30$, $c_{41} = 2.60$, $c_{42} = 3.00$.
Applying equations (2.18), (2.4) and (2.6), and noting that $cot^2\varphi_1 = cot^2\varphi_2 = 1/16$,

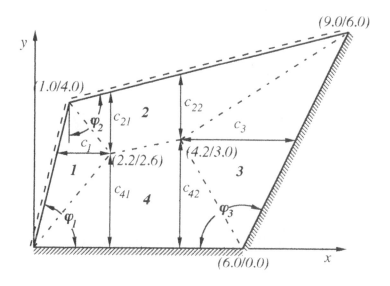

Fig. 7.2.1

$cot^2 \psi_3 = 1/4$ and $sin^2 \psi_3 = 0.8$, we get the following equilibrium equations for elements 1, 2, 3 and 4

$$m_{xf} + \frac{m_{yf}}{16} = \frac{9 \times 1.55^2}{6} = 3.60 \tag{7.1}$$

$$m_{yf} + \frac{m_{xf}}{16} = \frac{9}{6 \times 8.0} [1.7^2 \times 1.2 + (1.7^2 + 1.7 \times 1.8 + 1.8^2) 2.0 + 1.8^2 \times 4.8] =$$
$$= 7.01 \tag{7.2}$$

$$m_{xf} + \frac{m_{yf}}{4} - \frac{m_{s3}}{0.8} = \frac{9 \times 3.3^2}{6} = 16.34 \tag{7.3}$$

$$m_{yf} - m_{s4} = \frac{9}{6 \times 6.0} [2.6^2 \times 2.2 + (2.6^2 + 2.6 \times 3.0 + 3.0^2) 2.0 + 3.0^2 \times 1.8] =$$
$$= 19.55 \tag{7.4}$$

Solving these equations we find $m_{xf} = 3.17$, $m_{yf} = 6.81$, $m_{s3} = -9.17$, $m_{s4} = -12.74$.

In this case we find unique values of the average moments with a given pattern of lines of zero shear force. This is because there are two fixed edges and thus four moments to determine, equal to the number of equations. If there had been three fixed edges one of the moment values could have been chosen and if there had been four fixed edges two of the moment values could have been chosen. If, on the other hand, fewer than two edges are fixed the pattern of lines of zero shear force has to be arranged so that all four equations are satisfied although the number of unknowns is less than four.

A distribution of design moments is proposed in Fig. 7.2.2.

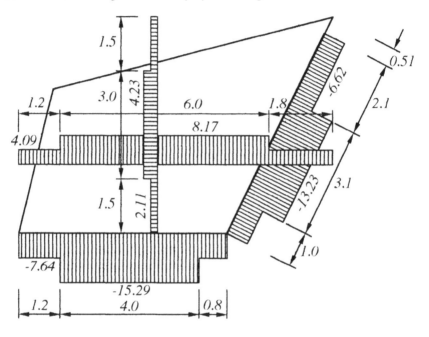

Fig. 7.2.2

7.2.2 One free edge

The application to non-rectangular slabs with one free edge is a mixture of the principles demonstrated for rectangular slabs with one free edge in Chapter 4 and the principles shown above for slabs with non-orthogonal edges.

Depending on the shape of the slab the choice of the main directions of bottom reinforcement and of the pattern of lines of zero shear force will differ. If the shape is very irregular, with all corners non-orthogonal, the numerical analysis becomes laborious, even if in principle it is no more difficult. Therefore only one example will be shown with a slab which is not too irregular.

Example 7.2

The slab in Fig. 7.2.3 has three fixed edges and one free edge. The fixed edges form right angles, but the free edge has a skew direction. The load is 12 kN/m².

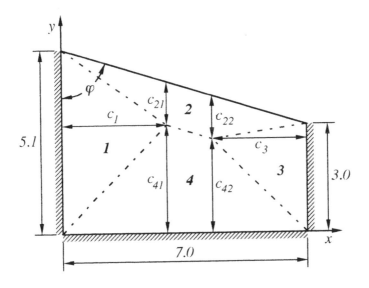

Fig. 7.2.3

The main bottom reinforcement directions are parallel to the x- and y-axes and the main support reinforcement is at right angles to the supports. A support band is also arranged along the free edge.

The analysis starts with the calculation of moments for the design of the main reinforcement. After that the design of the support band is carried out.

The figure shows a possible pattern of lines of zero shear force for the calculation of design moments for the main reinforcement. Other shapes of the pattern are also possible, e.g. with elements 2 and 4 as triangles and 1 and 3 as trapezoids, or with all four elements as triangles, meeting at a point. The c-values should be chosen so that the ratios between the different moments are estimated to be acceptable. A suitable choice of values can also simplify the numerical analyses. Thus in this example it is suitable to choose $c_{21} = c_{22}$. Of course the sums of c-values in the y-direction have to be adapted to the size of the slab.

The following numerical values have been used in the analysis below:

$c_1 = 3.0, c_{21} = c_{22} = 1.2, c_3 = 2.7, c_{41} = 3.0, c_{42} = 2.61$.

Applying equations (2.18), (2.4), (2.5) and (2.6) and noting that $\cot^2\varphi = 0.09$, we get the following equilibrium equations for elements 1, 2, 3 and 4

$$m_{xf} - m_{s1} = \frac{12 \times 3.0^2}{6} = 18.00 \tag{7.5}$$

$$m_{yf} + m_{xf} \times 0.09 = \frac{12 \times 1.2^2(7.0 + 2 \times 1.3)}{6 \times 7.0} = 3.95 \tag{7.6}$$

$$m_{xf} - m_{s3} = \frac{12 \times 2.7^2}{6} = 14.58 \tag{7.7}$$

$$m_{yf} - m_{s4} = \frac{12}{6 \times 7.0}[3.0^2 \times 3.0 + (3.0^2 + 3.0 \times 2.61 + 2.61^2)1.3 + 2.61^2 \times 2.7] =$$
$$= 21.75 \tag{7.8}$$

We have five unknown moments but only four equations. We thus have to choose one of the moment values. It is in this case not appropriate to choose a value for m_{yf} or m_{s4}, as the value of m_{xf} is very sensitive to such a choice. The best value to choose is m_{xf}. A suitable value may be $m_{xf} = 5.00$, which gives $m_{s1} = -13.00$, $m_{s3} = -9.58$, $m_{yf} = 3.50$, $m_{s4} = -18.25$.

The support band along the free edge has to carry the load from element 2. This load is assumed to be carried at right angles into the support band. The maximum load per unit length of the support band is $12 \times 1.2 \sin\varphi = 13.79$ kN/m. Fig. 7.2.4 shows the load on the support band. The moments in the band can be determined with usual methods. We will assume a point of zero shear force at a distance of 3.80 m from the left-hand end. With notation M_{bf} for the span moment and M_{bs1} and M_{bs3} for the support moments we get the following equilibrium equations:

$$M_{bf} - M_{bs1} = 13.79 \times 3.48^2/3 + 13.79 \times 0.32 \times 3.64 = 71.73 \tag{7.9}$$

$$M_{bf} - M_{bs3} = 13.79 \times 2.47^2/3 + 13.79 \times 1.04 \times 2.99 = 70.93 \tag{7.10}$$

If we choose $M_{bf} = 24.00$ we find $M_{bs1} = -47.73$, $M_{bs3} = -46.93$.

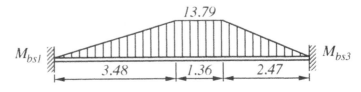

Fig. 7.2.4

According to the recommendations in Section 2.8.3 the bottom reinforcement in the support band should be distributed across a width of about half the average width of element 2.

This average width is 0.7 m. A width of 0.4 m may therefore be chosen, which gives 60 kNm/m as design moment for the bottom reinforcement.

The support moments in the support band correspond to a reinforcement direction which is parallel to the direction of the free edge. This direction is not at right angles to the supports of the slab. Support reinforcement is, however, most efficient if it is arranged at right angles to a support, as expected cracks will be parallel to the support. It is possible to take care of the moments M_{bs1} and M_{bs3} by means of reinforcement at right angles to the support. In this case the reinforcements only need to take the components $M_{bs1}\sin\varphi = -45.72$ and $M_{bs3}\sin\varphi = -44.95$ respectively. In this example we chose to take half the support moments in each way. Thus we reinforce for -23.87 and -23.47 respectively in the direction of the edge and add -22.86 and -22.48 respectively to the support moments for reinforcement at right angles to the supports. The reinforcement in the direction of the edge is concentrated in a width of 0.4 m whereas the reinforcement at right angles to the edge is less concentrated, see the recommendations in Section 2.8.3.

The total support moments to be taken by reinforcement at right angles to the edges are for support 1: $-13.00\times5.1 - 22.86 = -89.16$ kNm , and for support 3: $-9.58\times3.0 - 22.48 = -51.22$ kNm. These total moments have to be distributed with respect to their origins which means that the total moment contains one part from the element and one part from the free

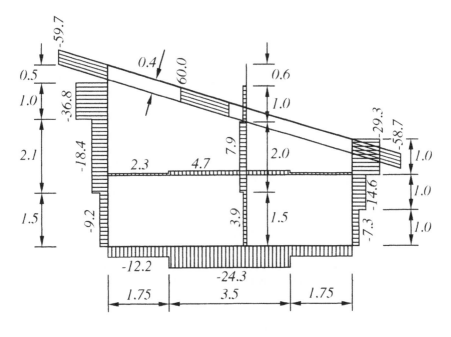

Fig. 7.2.5

edge; in approximate terms, the part from the band should be concentrated near the free edge. A suitable distribution of design moments is proposed in Fig. 7.2.5.

With the distribution shown, the concentration of reinforcement is higher at the upper end of the right-hand support than at the upper end of the left hand support, as two layers of support reinforcement are crossing in the former case. This seems to be in agreement with the moment distribution which can be expected according to the theory of elasticity.

7.2.3 Two opposite free edges

The load will in principle be carried by strips between the supported edges. This is mainly done by means of one-way strips, often with the aid of support bands along the free edges.

Example 7.3

One of the supports for the slab in Fig. 7.2.6 is fixed and one is a simple support. The load is $7\ kN/m^2$. The sizes of the slab are given by means of coordinates.

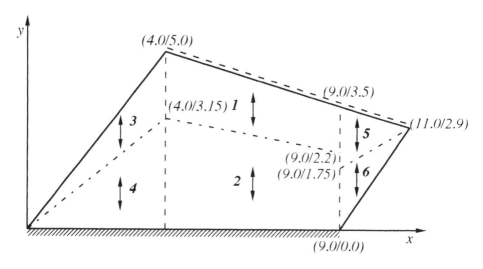

Fig. 7.2.6

The load is assumed to be carried by reinforcement which is at right angles to the fixed support. Theoretically it would be slightly better to arrange the span reinforcement at right angles to the dividing line between elements 1 and 2, the line of zero shear force, but this would complicate the analysis and the gain is small. The line of zero shear force between elements 1 and 2 is chosen approximately as the bisector between the support directions.

With the chosen shapes of the elements, which are shown in the figure, we get the average moments in elements 1 and 2 by means of Eq. (2.8)

$$m_{yf1} = \frac{7}{6}(1.85^2 + 1.85 \times 1.3 + 1.3^2) = 8.77 \qquad (7.11)$$

$$m_{yf1} - m_{s2} = \frac{7}{6}(3.15^2 + 3.15 \times 2.2 + 2.2^2) = 25.31 \qquad (7.12)$$

and thus $m_{s2} = -16.54$.

In the end parts of the slab triangular elements 3-6 have been chosen. Other possibilities exist, which may be more efficient, but are also more complicated. Eq. (2.4) gives for these triangular elements

$$m_{yf3} = 7 \times 1.85^2/6 = 3.99 \qquad (7.13)$$

$$m_{yf3} - m_{s4} = 7 \times 3.15^2/6 = 11.58 \qquad (7.14)$$

and thus $m_{s4} = -7.59$.

$$m_{sf5} = 7 \times 1.75^2/6 = 3.57 \qquad (7.15)$$

Elements 3 and 6 are supported on support bands along the free edges. As the elements are of a pure one-way type, the load is directly carried in the y-direction and gives triangular load distributions on the bands with the maximum load at one end. For the calculation of moments in the support bands, ordinary formulas for beams according to the theory of elasticity can be used in this case.

The support band along element 3 carries a total load of $7 \times 1.85 \times 4.0/2 = 25.90$ kN and has a span of 6.40 m. This gives a support moment of $-0.1167 \times 25.90 \times 6.40 = -19.34$ and a span moment of $0.0846 \times 25.90 \times 6.40 = 14.02$ kNm. The average width of the element is about 0.6 m and according to the recommendations in Section 2.8.3 the span reinforcement may be distributed over a width of about 0.4 m, as this width is not of great importance for the behaviour of the slab.

The support band along element 6 carries a total load of $7 \times 1.75 \times 2.0/2 = 12.25$ kN and has a span of 3.52 m. This gives a support moment of $-0.1333 \times 12.25 \times 3.52 = -5.75$ and a span moment of $0.0596 \times 12.25 \times 3.52 = 2.57$ kNm. The average width of the element is about 0.5 m and a corresponding distribution width of the span reinforcement is about 0.4 m.

A distribution of design moments based on the calculated values is proposed in Fig. 7.2.7. The main condition is that the total design moment in a section has to correspond to the sum of the calculated moments. Thus, for example, the sum of span moments for the reinforcement in the y-direction is $4.0 \times 3.99 + 5.0 \times 8.77 + 2.0 \times 3.57 = 66.9$ kNm. The lateral distribution is chosen with respect to the variation of the distance between the support and the line of zero shear force.

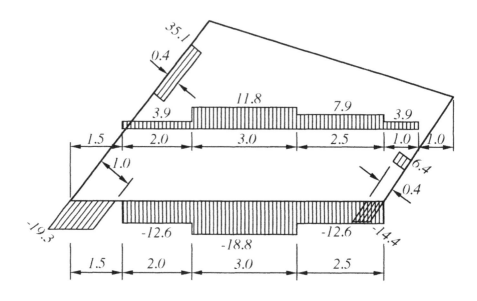

Fig. 7.2.7

At the lower right-hand corner two layers of top reinforcement have been proposed which cross over each other, one layer at right angles to the support and one layer along the free edge. The choice of this arrangement is influenced by the fact that a moment concentration can be expected at such a corner according to the theory of elasticity. The concentration of reinforcement prevents the formation of large cracks at this position.

7.2.4 Two adjacent free edges

It has been demonstrated in Section 5.2 that the design of a rectangular slab with two adjacent free edges may be complicated. Of course the design of a slab with non-orthogonal edges and two adjacent free edges is still more complicated. Solutions with the types of element which are used in most of the other cases cannot be applied in this case. We therefore have to make direct use of the basic principles of the strip method with support bands in one direction supporting strips in other directions, as described in Section 2.8.

Example 7.4

The slab in Fig. 7.2.8 carries a load of 7 kN/m^2. It has one fixed edge *AB*, one freely supported edge *AD* and two free edges *BC* and *CD*.

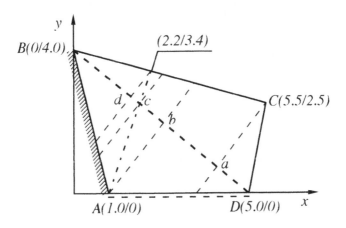

Fig. 7.2.8

A support band is assumed to have its supports at *B* and *D*. According to the rules in Section 2.8 the support band is assumed to have zero width when it acts as support for crossing strips. The support band follows a line with the equation $y = 0.8(5 - x)$. The crossing strips, which are supported by the support band, are assumed to have a direction at right angles to the support band. This is often the best assumption from the points of view of reinforcement economy and behaviour under service conditions. There may however be situations where another direction is preferable. This will be discussed at the end of this example.

As one support is fixed, some top reinforcement should also be arranged at right angles to that support. This reinforcement can be assumed to carry all the load within a certain distance from the support. It is assumed that it carries all the load to the left of a line from *A* to the point on the edge with coordinates (2.2/3.4). The choice is made such that this support moment is of approximately the same magnitude as the negative moment above the support band, which seems reasonable in this case. The line cuts the support band at point *c* with coordinates (1.88/2.50). The distance from (2.2/3.4) to the edge *AB* is 1.99 m. Application of Eq. (2.4) gives an average support moment

$$m_{sAB} = -\frac{7 \times 1.99^2}{6} = -4.62 \qquad (7.16)$$

In the remaining part of the slab the load is carried by strips at right angles to the support band. Some typical strips are now considered.

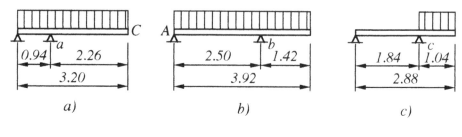

a)   b)   c)

Fig. 7.2.9

The strip through point C is a line with the equation $y=1.25(x-3.5)$, which cuts the support band at point a with coordinates (4.09/0.73). The strip and its load are shown in Fig. 7.2.9 a). The support reactions acting at the support band at point a and at the supported edge are

$$R_a = \frac{7 \times 3.20^2}{2 \times 0.94} = 38.1 \tag{7.17}$$

$$R_{edge} = 7 \times 3.20 - 37.7 = -15.7 \tag{7.18}$$

There is thus a negative reaction at point (3.5/0), which means that the slab has to be anchored.

In the same way we can analyse a strip through point A, which has the equation $y=1.25(x-1.0)$. This strip cuts the support band at point b with coordinates (2.56/1.95) and the edge at (3.45/3.06). The strip and its load are shown in Fig. 7.2.9 b). We find the support reactions

$$R_b = \frac{7 \times 3.92^2}{2 \times 2.50} = 21.5 \tag{7.19}$$

$$R_{edge} = 7 \times 3.92 - 21.5 = 5.9 \tag{7.20}$$

We can continue to choose the strip through point c, which has the equation $y - 2.50 = 1.25(x-1.88)$. This line cuts the free edge at (2.53/3.31) and the fixed edge at (0.73/1.06). The strip and its load is shown in Fig. 7.2.9 c). For simplicity it is assumed that there is a free support at line AB. We find the reaction at c

$$R_c = \frac{7 \times 1.04 \left(1.84 + \frac{1.04}{2}\right)}{1.84} = 9.3 \qquad (7.21)$$

At last we look at the strip through (2.2/3.4). This strip cuts the support band at (1.63/2.69). At this point the load on the support band has fallen to zero.

Fig. 7.2.10 illustrates the load on the support band. Some additional points have been calculated in the same way as above in order to be able to draw the curve. On the other hand a quite satisfactory result would have been achieved with straight lines between the calculated values above.

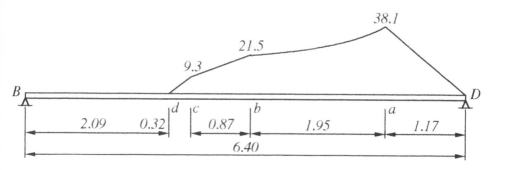

Fig. 7.2.10

An analysis of the support band as a simply supported beam with the load according to Fig. 7.2.10 gives the support reactions $R_B = 29.1$ and $R_D = 64.4$. The maximum moment in the support band is $M_b = 93.5$ at a distance of 3.9 m from B and 2.5 m from D.

In order to determine a suitable width for the reinforcement band the recommendations in Section 2.8.3 are used. The average width of the parts of the slab which are supported by the support band is not well defined in this case, but it may be estimated to be about 2.5 m. As the support band in this case is of great importance for the behaviour of the slab a suitable width is $2.5/2 = 1.25$ m. With this width we find the maximum moment

$$m_b = 93.5/1.25 = 74.8 \qquad (7.22)$$

As the slab is cantilevering outside the support band some minimum reinforcement for a negative moment $m_\perp$ must be introduced at the ends of the support band according to Eq. (2.36). At D the angles according to Fig. 2.8.2 are $\varphi_1 = 62.7°$, $\varphi_2 = 38.7°$ and we get

$$\left|m_\perp\right| \geq \frac{1}{74.8} \left(\frac{2 \times 64.4}{\pi} \times \frac{62.7}{62.7 + 38.7}\right)^2 = 8.6 \tag{7.23}$$

This is a negative moment, requiring top reinforcement within a distance equal to $2.5/3 = 0.83$ m from D.

We also have to make the same analysis for the support at B. There we find $\varphi_1 = 23.4°$ and $\varphi_2 = 37.3°$ and

$$\left|m_\perp\right| \geq \frac{1}{74.8} \left(\frac{2 \times 29.1}{\pi} \times \frac{23.4}{23.4 + 37.3}\right)^2 = 0.7 \tag{7.24}$$

The bending moments in the strips through a, b and c can easily be calculated from Fig. 7.2.9. The results are 17.9, 7.1 and 3.8, respectively. The total moment in the strips resting on the support band is most easily calculated by means of Eq. (2.6), using numbers from Figs 7.2.9 and 7.2.10 and noting that the distance from d to $(2.2/3.4)$ is 0.91 m.

$$M = -\frac{7}{6}[2.26^2 \times 1.17 + (2.26^2 + 2.26 \times 0.91 + 0.91^2)\,3.14 - 0.91^2 \times 0.32] =$$

$$= -35.9 \tag{7.25}$$

Fig. 7.2.11 shows a proposed distribution of negative design moments. All the reinforcement at right angles to the support band should be carried to the edges. The reinforcement for the support moment -3.1 close to corner B can take a moment across the support band equal to $-3.1\sin^2 37.3° = -1.1$, where $37.3°$ is the angle between edge AB and the support band. Thus condition (7.24) is fulfilled.

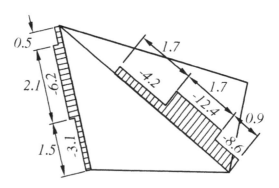

Fig. 7.2.11

The support band should be designed for a positive moment $m_b = 72.4$ across a width of 1.25 m symmetrical to the support band. All this reinforcement should be carried to the edges and be well anchored. In addition to the band reinforcement, minimum bottom reinforcement should be introduced, preferably parallel to the band, and possibly also some bottom reinforcement at right angles.

As the reinforcement in the support band is so much heavier than the rest of the reinforcement it may be questioned whether this is a suitable design. Could we get a more even reinforcement distribution by using some other design method? If, instead, we apply the theory of elasticity in a theoretically correct way (e.g. using elements which are small enough in the finite element method), we would presumably get a still higher maximum moment close to corner D. To apply the yield line theory to this slab in a correct way is extremely difficult, as checks of many different yield line patterns would have to be carried out, including fan-shaped local patterns, for instance in the vicinity of corner D.

In fact it is questionable whether this is a suitable way of making the slab, or whether it would not be better to use a real beam instead of a support band.

Instead of arranging the top reinforcement at right angles to the support band it may be arranged at another angle. If, for example, a slope of 1/1.65 is used instead of 1/1.25 there will be no negative reaction along edge AB and the load on the support band will be lower. This would lead to smaller moments in the support band and a corresponding reduction of reinforcement. On the other hand the reinforcement crossing the band will increase. Depending on the angles between the edges of the slab this type of solution may sometimes be the most suitable.

7.3 Other cases

7.3.1 Circular slabs with a uniform load

Circular slabs with polar symmetric load and support can easily be analysed with the equilibrium equation expressed in polar coordinates. Such analyses are shown in *Strip Method of Design*, Chapter 3. The results are presented as radial and tangential moments. In practice it is easier to arrange the bottom reinforcement as a rectangular mesh of in two directions at right angles to each other. It can be demonstrated that the bottom reinforcement in both directions should then be designed for the tangential moment.

For a simply supported circular slab with radius r and a uniform load q, reinforced in two directions at right angles, the average moments according to this analysis are

$$m_x = m_y = \frac{qr^2}{6} \tag{7.26}$$

The corresponding reinforcement can be evenly distributed and should in that case be carried to the support. It may, however, be better to concentrate it somewhat towards the centre, e.g. as shown in Fig. 7.3.1.

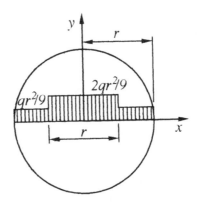

Fig. 7.3.1

The slab can also be treated by direct use of the simple strip method according to Fig. 7.3.2, where the load is carried to the nearest support in one of the two reinforcement directions. As the support is curved the average moment is most simply calculated by means of a numerical integration. The result shows an average moment which is about 16% lower than the value above, and thus theoretically more economical. The corresponding moment distribution is also shown in Fig. 7.3.2 and compared to a solution according to the theory of elasticity (curve marked *el.*). It can be seen that, in this case, the moment distribution determined from the direct use of the simple strip method is not very suitable with respect to service conditions, as it does not give any design reinforcement across the outer part of the diameter. The distribution of reaction force along the edge is very uneven according to the simple strip solution with zero shear force in the 45° direction. This also shows that the solution is far from the reality of the service state. It is therefore recommended that values should be chosen according to the equation above and a distribution of design moments which is approximately according to Fig. 7.3.1 or is constant.

If the edges of the slab are fixed the support reinforcement is best arranged at right angles to the support, i. e. in a radial arrangement. In the normal case the top reinforcement is thus radial and the bottom reinforcement is an orthogonal net.

If the support moment is m_s the following relation for the average design moments is valid according to the polar symmetric solution:

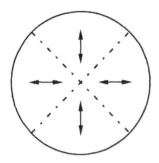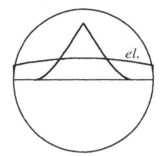

Fig. 7.3.2

$$m_{xf} - m_s = m_{yf} - m_s = \frac{qr^2}{6} \tag{7.27}$$

It may be noted that this value is the same as for a circumscribed square, see Fig. 7.3.3, where c for the elements in the square corresponds to r for the circle. The valid formula for the elements is (2.4). Thus the average moments in a circular slab can be taken as the moments in a circumscribed square slab with the sides parallel to the directions of the bottom reinforcement.

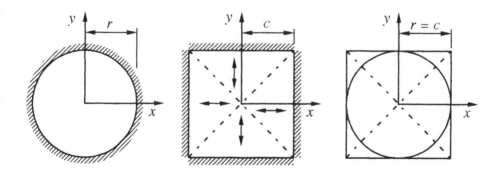

Fig. 7.3.3

7.3.2 General case with all edges supported

Where a slab is supported around all its edges it is always possible to apply the simple strip method, dividing the slab into narrow strips in the reinforcement directions, although the numerical computations may become lengthy. Also the result may be a design which is not appropriate for the behaviour in the service state, as the design moment may be zero within some parts where appreciable moments occur under service conditions. Some minimum reinforcement then has to be added.

One way of simplifying the analysis may be to apply the following general rule:

It is always safe to design the slab for the moments in a circumscribed slab with the same bottom reinforcement directions and with all support reinforcement at right angles to the supports in both slabs.

If this rule is applied, the reinforcement can be curtailed according to the rules given in Section 2.9.1, provided that the distance to the support is taken as the distance to the edge of the circumscribed slab for the bottom reinforcement and to the edge of the real slab for the support reinforcement. The c-value is taken from the circumscribed slab.

Example 7.5

The elliptical slab in the upper part of Fig. 7.3.4 has fixed edges and carries a load of 9 kN/m². The design moments are first determined from the circumscribed rectangular slab in the lower part of the figure. The choice of lines of zero shear force is shown. Based on these lines and Eqs. (2.5) and (2.4) the average design moments are:

$$m_{xf} - m_{xs} = \frac{9 \times 1.5^2}{6} = 3.38 \tag{7.28}$$

$$m_{yf} - m_{ys} = \frac{9 \times 2.0^2 (8.0 + 2 \times 5.0)}{6 \times 8.0} = 13.50 \tag{7.29}$$

We can choose $m_{xs} = -2.25, m_{xf} = 1.13, m_{ys} = -9.00, m_{yf} = 4.50$. A distribution of design moments based on these values is proposed in Fig. 7.3.5. The limit between the parts where m_{xs} and m_{ys} are active is where the corresponding line of zero shear force cuts the ellipse in Fig. 7.3.4. The distribution of m_{ys} might have been chosen with some concentration towards the centre, as the support moments in the service state are highest there.

This way of using the moments from the circumscribed rectangle gives very simple calculations and a conservative result. On the other hand the design may be regarded as too conservative and thus uneconomical. In order to get a more economical design a direct application of the strip method is possible, dividing the slab into narrow strips in the x-and y-directions for span moments and cantilevering strips at right angles to the support for the support moments. Where the span strips rest on the ends of the cantilevers due regard has to

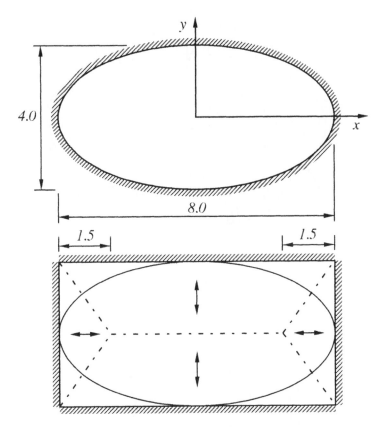

Fig. 7.3.4

be given to the difference in directions according to the rules in Section 2.3.6. Such an analysis is lengthy and time-consuming and will not be shown here.

However, there also exists the possibility of circumscribing the ellipse with a polygon instead of a rectangle. This gives results which are not so conservative and yet the calculations are not too lengthy or complicated. Fig. 7.3.6 shows the ellipse circumscribed by a polygon shaped by introducing lines with the slopes ±0.5, which are tangents to the ellipse. Only one quarter of the slab is shown, as it is symmetrical with respect to both the x- and y-axes.

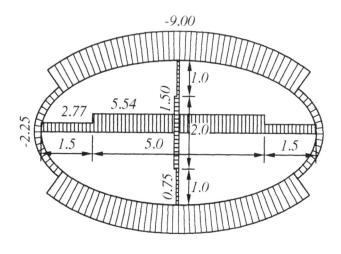

Fig. 7.3.5

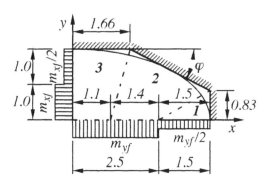

Fig. 7.3.6

The assumed lines of zero shear force are shown in the figure, as well as the assumed distribution of span design moments. Note that m_{xf} and m_{yf} in this case do not mean average moments but moments in the relevant parts.

For elements 1 and 3 we get, by means of Eqs (2.4) and (2.5),

$$m_{xf} - m_{s1} = \frac{9 \times 1.5^2}{6} = 3.38 \qquad (7.30)$$

$$m_{yf} - m_{s3} = \frac{9 \times 2.0^2 (1.66 + 2 \times 1.1)}{6 \times 1.66} = 13.95 \qquad (7.31)$$

For element 2 we can apply Eq. (2.18) together with (2.6). The direction of edge 2 cuts the x-axis at 5.66. The following values are introduced into Eq. (2.6):

$l = 1.17; l_1 = 2.0; l_2 = 0; l_3 = -0.83; c_1 = 4.56; c_2 = 3.16;$

We also have to introduce the average span moments on the width corresponding to edge 2. This gives us the following relation:

$$\frac{0.17 + \frac{1.0}{2}}{1.17} m_{xf} + \frac{0.84 + \frac{1.5}{2}}{2.34} m_{yf} \cot^2\varphi - \frac{m_{s2}}{\sin^2\varphi} =$$

$$\frac{9}{6 \times 1.17} (4.56^2 \times 2.0 - 3.16^2 \times 0.83) = 42.69 \qquad (7.32)$$

Introducing $\cot^2\varphi = 4$, $\sin^2\varphi = 0.2$, the relation can be written

$$0.573 m_{xf} + 2.718 m_{yf} - 5 m_{s2} = 42.69 \qquad (7.33)$$

If we choose $m_{xf} = 1.13$ and $m_{yf} = 4.65$ we get $m_{s1} = -2.25$, $m_{s2} = -5.88$, $m_{s3} = -9.30$. The corresponding distribution of design moments is shown in Fig. 7.3.7.

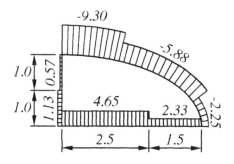

Fig. 7.3.7

133

This analysis reduces the total amount of reinforcement by about 14%, compared with the analysis based on a circumscribed rectangle. It also gives a more satisfactory reinforcement distribution

The distribution of support moments and the ratio between moments in the x- and y-directions can be changed by changing the assumed points where the lines of zero shear force meet the x-axis.

Example 7.6

The slab in Fig. 7.3.8 has three fixed and two freely supported edges and a uniform load of 7 kN/m^2. It is to be reinforced with bottom reinforcement in the directions of the coordinate axes and support reinforcement at right angles to the supports.

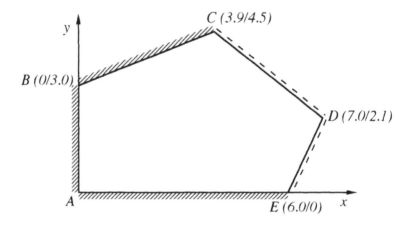

Fig. 7.3.8

Three different ways of performing the design will be discussed.

The first method is to use the moments from the circumscribed rectangle. This method is by far the simplest, but it may lead to a very coservative , i.e. uneconomical, solution.

The circumscribed rectangular slab is shown in Fig. 7.3.9. In order that this slab shall correspond to the real slab as far as possible the two long edges are assumed to be fixed only on the parts adjacent to the fixed edges in the real slab. Assumed lines of zero shear force are also shown in the figure. Based on these lines we get the following relations from Eqs. (2.4) and (2.5). The moments m_s are the average moments on the fixed parts of the edges.

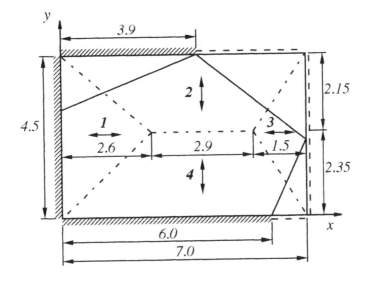

Fig. 7.3.9

$$m_{xf} - m_{s1} = \frac{7 \times 2.6^2}{6} = 7.89 \tag{7.34}$$

$$m_{xf} = \frac{7 \times 1.5^2}{6} = 2.63 \tag{7.35}$$

$$m_{yf} - \frac{3.9}{7.0} m_{s2} = \frac{7 \times 2.15^2 (7.0 + 2 \times 2.9)}{6 \times 7.0} = 9.86 \tag{7.36}$$

$$m_{yf} - \frac{6.0}{7.0} m_{s4} = \frac{7 \times 2.35^2 (7.0 + 2 \times 2.9)}{6 \times 7.0} = 11.78 \tag{7.37}$$

We get $m_{xf} = 2.63$, $m_{s1} = -5.26$ and if we choose $m_{yf} = 4.50$ we find $m_{s2} = -9.62$ and $m_{s4} = -8.49$.

Support *BC* in the real slab corresponds partly to edge 1 and partly to edge 2 in the circumscribed rectangular slab. The corresponding parts depend on where the line of zero shear force cuts the edge. We then find that 32% corresponds to edge 1 and 68% to edge 2. If we use these percentages we find the average support moment -8.22.

As usual a suitable distribution of the different average moments should be chosen.

135

This simple design is on the safe side, but probably too much so. Therefore an alternative, but more laborious, solution will be demonstrated, based on the analysis of equilibrium conditions for elements, each of which has one side along an edge of the slab. A configuration of lines of zero shear force, which divides the slab into such elements, is shown in Fig. 7.3.10.

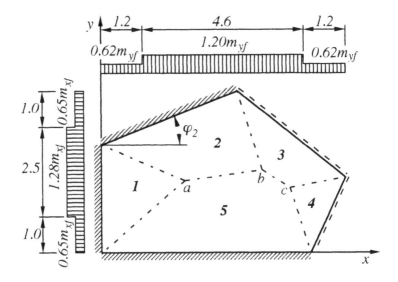

Fig. 7.3.10

In determining the pattern of lines of zero shear force some general rules may be followed. The elements at the shorter edges are usually triangles. The lines between the elements at the longer sides run approximately in a direction towards the point where the directions of these edges meet. In this way the general shape may be determined. For the complete determination of the shape a trial and error procedure has to be used in order to get suitable relations between the different moments. In this procedure the positions of points a, b and c are therefore varied. Often it is sufficient to exactly fulfil the equilibrium conditions only for the largest elements but a check should be made that the design moments are large enough to maintain equilibrium in the smaller elements. This is on the safe side, but with only a small influence on the economy of the design.

Before starting the equilibrium analyses the distribution of the span design moments should be decided, as this distribution influences the equilibrium conditions. An assumed distribution has been given in the figure.

In this case suitable choices are the following coordinates for the points:
$a(2.4/2.0)$, $b(4.6/2.3)$, $c(5.4/1.8)$

The complete analysis will not be shown, as it is rather laborious, but as an example the equilibrium condition for element 2 will be given. It is based on Eq. (2.18) combined with (2.6). The geometric values to be used in Eq. (2.6) are the following:

$l = 1.5, l_1 = 2.2, l_2 = 0.3, l_3 = -1.0, c_1 = 6.42, c_2 = 5.0.$

Using these values and the relevant average span moments we get the equilibrium condition for element 2

$$\frac{1.0 \times 0.65 + 0.5 \times 1.28}{1.5} m_{xf} + \frac{1.2 \times 0.62 + 2.7 \times 1.20}{3.9} m_{yf} \cot^2 \varphi - \frac{m_{s2}}{\sin^2 \varphi} =$$

$$= \frac{7}{6 \times 1.5}[6.42^2 \times 2.2 + (6.42^2 + 6.42 \times 5.0 + 5.0^2) 0.3 - 5.0^2 \times 1.0] = 74.02 \qquad (7.38)$$

or, with $\cot^2 \varphi_2 = 6.76$, $1/\sin^2 \varphi_2 = 7.76$

$$0.860\, m_{xf} + 6.906\, m_{yf} - 7.760\, m_{s2} = 74.02 \qquad (7.39)$$

Writing the corresponding equations for the other four elements and choosing values of the span moments we can find the following moment values:

$m_{xf} = 2.19, m_{yf} = 3.26, m_{s1} = -4.38, m_{s2} = -6.40, m_{s5} = -6.63$

When we compare these values with the previous solution we find that they are on average about 20% lower. We can thus, in this case, save about 20% reinforcement by using this more complicated solution. This comparison must be take into account the fact that the latter solution may involve several hours work before a suitable pattern of lines of zero shear force is found from the trial and error process, whereas the first solution can be made in a couple of minutes. The cost of reinforcement has to be compared to the cost of time.

A third possibility is a direct application of the principles of the simple strip method. Fig. 7.3.11 illustrates how this may be performed. The slab is divided into zones with different load-bearing directions. The dashed lines are in this case lines of zero moments. The zones by the fixed supports are zones where the support reinforcement is active. These support strips act as cantilevers with zero moment at the ends. In the remaining parts the load is carried by span strips parallel to the bottom reinforcement, in the direction of the coordinate axes. The load within a certain part may also be divided between the two reinforcement directions. The span strips rest on the cantilevers and on the supports. The moments are calculated in narrow strips at suitable distances, which gives a moment distribution and an average moment of each type. A suitable distribution of the average moments is used as the basis for design. The ratios between different average moments depend on how the limits are chosen between different parts.

Where a span strip rests on a support strip due regard has to be paid to the change in direction according to the rules in Section 2.3.6.

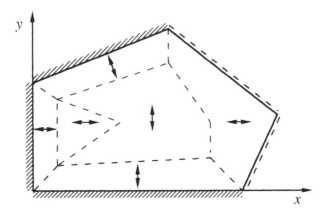

Fig. 7.3.11

With the strip directions in Fig. 7.3.11 there will be no design reinforcement in the y-direction near the right-hand side of the slab. Some minimum reinforcement can be arranged there, but it is also possible to get design reinforcement by dividing the load in this region between the two directions.

This type of analysis is rather time-consuming, as it takes time to analyse a number of strips. The analysis may also have to be repeated with other limits between the different zones, if the ratios between different moments prove to be unsuitable.

7.3.3 General case with one straight free edge

Where one straight edge is free a support band has to be assumed along this edge. The analysis may be based on a combination of the second approach in the example above and the method demonstrated in Section 7.2.2. As an alternative, a direct application of the simple strip method may be used as discussed above and demonstrated in Section 7.2.4.

7.3.4 General case with two or more free edges

This complicated case has to be analysed by means of the direct use of the simple strip method, including support bands, in principle as demonstrated in Section 7.2.4.

CHAPTER 8

Regular flat slabs with uniform loads

8.1 General

8.1.1 Definition of "regular"

In this chapter the word *regular* means that the supports form an orthogonal net and that all interior supports are columns. The exterior supports may be walls or columns and the slab may cantilever outside the exterior supports. There should be no major openings in the slab.

The *advanced strip method* was developed in the first place for the design of the regular flat slabs treated in this chapter. Provided that certain approximations are accepted it can, however, be applied to a much wider group of slabs, treated in other chapters.

8.1.2 Drop panels and column capitals

A drop panel, Fig. 8.1.1, means that the slab is given a greater depth in an area around a column. In this way the slab becomes stronger in the part where the moments and shear forces are largest, without much increase in the moments caused by the dead weight.

Where drop panels are used, the widths of the column strips are best chosen equal to the widths of the drop panels, at least regarding support reinforcement at the columns. The widths of the column strips may be chosen larger for the span reinforcement.

As a drop panel increases the stiffness of the slab at the support it may be appropriate to increase the support moment somewhat compared to the case without a drop panel.

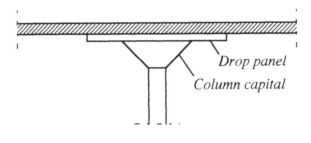

Fig. 8.1.1

A column capital, Fig. 8.1.1, is intended to increase the punching strength and at the same time decrease the bending moments in the slab.

A column capital decreases the ratio between support and span moments according to the theory of elasticity, compared to the ratio without a capital.

8.1.3 Determination of span

The main type of element in a flat slab is the corner-supported element, supported on the circumference of a column or a column capital. Theoretically the support area is infinitely small which gives an infinitely high stress. In practice, of course, the support area and support stress have to be finite. This may be interpreted so that the reaction force acts somewhere inside the column and not at some point at the circumference.

If we look at the ultimate limit state, a crack will appear at the circumference of a concrete column, where thus the critical section is. For the determination of design moments the span can therefore be assumed to be taken to the circumference of the column, and the theoretical point of support for the corner-supported element at that circumference. If there is a sufficiently strong column capital the point of support may be assumed to be at its circumference.

If we accept that a point on the circumference is chosen as the theoretical point of support, from which the span is calculated, the same point is, of course, valid for both directions of the element. Thus, if we have circular columns with radius r at equal centres l_c in both directions, the spans may be taken as $l = l_c - r\sqrt{2}$, provided that the points of support are taken in the 45°-direction from the column centres, see Figs 8.1.2 and 8.1.3. By changing the assumed point of support the span can be somewhat decreased in one direction but then it has to be increased in the other direction.

It is important to note that the span may *not* be taken as the clear distance $l-2r$, which is used in some codes. It can easily be demonstrated by means of the yield line theory that such an assumption is theoretically unsafe.

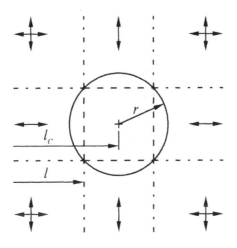

Fig. 8.1.2

In analysis by the strip method there is a one-way strip between, and acting together with, the corner-supported elements, see Fig. 8.1.2. Where the columns are circular the span of such a one-way strip is a little shorter than that of the corner-supported elements, closer to l-$2r$. As the one-way strips are of a minor importance for the total amount of reinforcement, their spans can be taken to coincide with those of the corner-supported elements.

In summary, the following rule is used for the determination of the span where the slab is supported on a concrete column:

The span of a strip resting on a concrete column is measured from the side of an inscribed rectangle with edges parallel to the reinforcement.

If the column is made of some other material, e.g. masonry, the span has to be made a little longer.

8.1.4 Calculation of average design moments

The elements are combined to form strips in the way which is exemplified in Fig. 8.1.3. The whole width w_x can be treated as one wide strip in the x-direction and the whole width w_y as one wide strip in the y-direction. The whole static analysis is limited to the analysis of these two strips in the same way as a continuous beam is analysed. This analysis gives the average design moments.

When the average design moments are known, the positions of the lines of zero shear force (maximum moments) can be calculated. Thus the moments in the x-direction determine the widths of the individual strips and elements in the y-direction and vice versa.

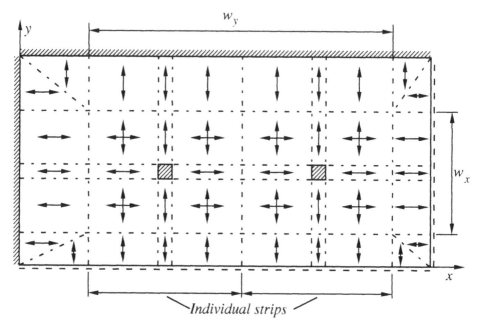

Fig. 8.1.3

The determination of support moments in the strips can be based on the theory of elasticity and calculated by ordinary methods for continuous beams. It is, however, not necessary to calculate the support moments accurately according to the theory of elasticity, although the ratio between support and span moments should not deviate too much from that derived from the theory. Sufficient agreement with the theory of elasticity is achieved if the following recommendations are followed. See also Section 1.5.

First support moments are calculated for each span according to the theory of elasticity assuming that it is fixed at interior column supports, i.e. $ql^2/12$ if the opposite end is fixed and $ql^2/8$ if the opposite end is freely supported. If the opposite end is continuous with a wall an intermediate value may be used, e.g. $ql^2/10$. If there is a drop panel which increases the stiffness of the slab over the column the support moments may be increased. If the support for the slab is wide, e.g. where a column capital is used, the support moments may be decreased. The design support moment is taken as the average of the moments from the spans meeting at the support, provided that no moment is assumed to be taken by the column. Starting from this average value some modification can be made. If it is important to minimize cracks on the top side of the slab the support moment can be somewhat increased and if it is important to minimize cracks on the bottom side the support moment can be somewhat decreased. Such a change should not be greater than about 15%. Within the limits given by these recommendations the influence of the ratio between support and span moments has no noticeable influence on deflections in the service state.

Regard should be given to moment transfer between the slab and the columns, where this transfer may be of importance for the behavior of the structure. For interior columns the moment transfer may often be disregarded, but for exterior columns this is as a rule not to be recommended. How these moments are calculated is outside the scope of the strip method. For the sake of simplicity they are therefore disregarded in the examples, with the exception of Example 8.4. This does not mean that it is generally recommended to disregard the moments. Relevant code rules have to be followed.

8.1.5 Lateral distribution of reinforcement

The strip method in itself only gives some limits for the distribution of design moments across the width of corner-supported elements, see Section 2.5.2. Some additional rules may be needed for crack control under service conditions.

The simplest possible moment distribution is with a constant span moment over the whole width and the support moment only within a limited width (the *column strip*) over the column, leaving the *middle strip* between the columns without top reinforcement. This choice leads to simple drawings and simple construction and is thus favourable from the point of view of economy. It is also suitable for limitation of deflections. It is, however, not the best design for crack limitation. It may lead to visually unacceptable top cracks between the columns and is only recommended where such cracks will be covered by some floor finish.

For the best possible crack control the design moment (reinforcement) should be distributed with some regard to the moment distribution according to the theory of elasticity. Where crack control on the top surface is essential the design support moment in the middle strip should be chosen to be 30-50% of the average support moment. Where crack control is essential on the bottom surface the design span moment in the column strip may be increased by about 20%. In both cases the average moments are kept unchanged.

The theoretical moments are different in the corner-supported elements and in the one-way strip between them, as there is a uniform moment in the latter element. When lateral moment distributions are discussed half the one-way strip is assumed to belong to each of the corner-supported elements in order to simplify the discussion.

The width of the column strip may generally be chosen as half the total width of the individual strip. (An individual strip is a strip with a width equal to the sum of the widths of two corner-supported elements supported on the same column and the one-way strip in between, see Fig. 8.1.3). Where there is a difference in span between the x- and y-directions it may be appropriate to choose a smaller width for the column strip in the shorter direction, e.g. so that it has the same width in both directions. The width of the column strip should never be chosen larger than half the total width of the individual strip.

Where a moment transfer between a column and a slab is taken into account in the design, the reinforcement has to be arranged in such a way that this moment transfer can take place.

In the vicinity of an exterior column there should be some torsional reinforcement along the free edge of the slab in order to limit torsional cracking.

Typical recommended lateral moment distributions are illustrated in Fig. 8.1.4. Distribution *s1* is the normally recommended distribution of the support moment where limitation of cracks on the top surface is not important, whereas *s2* shows an example of a suitable distribution for crack limitation on the top surface. Distributions *s3* and *s4* are examples of corresponding distributions of support moments in the shorter direction where the spans are different in the two directions. Distribution *f1* is the normally recommended even distribution of span design moments, whereas *f2* may be used where crack limitation on the bottom surface is particularly important.

In Fig. 8.1.4 the column strip is shown with the eccentricity which results from the application of the corner-supported element. In practice the reinforcement may be placed centrally over an interior column.

The chosen distributions should be checked to ensure that they fulfil the condition in Eq. (2.24). This condition is automatically fulfilled if distribution *s1* or *s2* or an intermediate between them is used for the support moment and *f1* for the span moment, and the ratio between support and span moment is between -1 and -3, which is the most common case.

Where there is no support moment, the span moment must be more concentrated in order to fulfil the condition according to Eq. (2.24), which means that the moment in the middle strip must not be larger than $0.7m_f$. This may be the case where there is no moment transfer from an exterior column.

Where there is no span moment, i.e. where a slab is cantilevering outside an exterior column, the support moment must be distributed so that the condition according to Eq. (2.24) is fulfilled. Distributions *s2* and *s4* are on the safe side with respect to this condition.

8.1.6 Summary of the design procedure

The determination of the design moments according to the advanced strip method is made in the following steps:

1. Determine the length of the spans as described in 8.1.3.

2. Determine the average support moments as described in 8.1.4.

3. Calculate the *c*-values by means of Eq. (2.34) and, from these, calculate the widths of the individual strips.

4. Calculate the average span moments by means of Eq. (2.35).

5. Choose lateral moment distributions according to the recommendations in 8.1.5. Check against Eq. (2.24) if this is thought to be necessary.

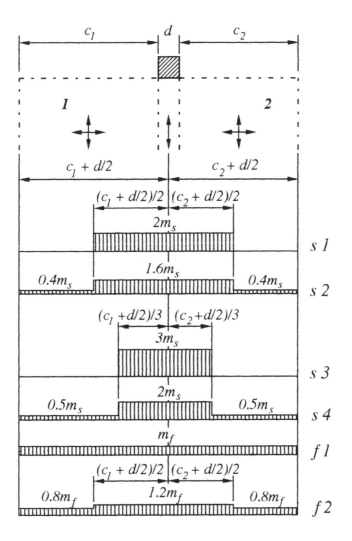

Fig. 8.1.4

6. Use these design moments for the design of the reinforcement. The design moments in the triangular elements in the corner of the slab (see Fig. 8.1.3) are taken as one third of the moments in the parallel parts of the strips. Where there are exterior columns arrange torsional reinforcement at the free edge.

7. Determine the length of reinforcing bars according to the rules in Section 2.10.

8. Calculate the support reactions at the columns and check for punching according to rules in handbooks or codes. The support reaction is equal to the load inside the lines of zero shear force in the spans surrounding the column.

The examples will generally only show the application of points 1-4 above, as the choice of lateral moment distribution depends on the requirement for crack control in the actual situation.

8.2 Exterior wall or beam supports

8.2.1 One single interior column

Whereas many methods for the design of flat slabs are only valid if there is at least a certain number of spans in each direction, the strip method is not limited by any such rule. It may thus be applied to a slab which is supported on walls or beams all around and on one single interior column. This case is easily treated with the strip method although it is rather complicated with other methods, such as the yield line theory, at least if the slab is not symmetrical with respect to the column.

Example 8.1

The slab in Fig. 8.2.1 carries a uniform load of 8 kN/m^2. Support A can be assumed to be fixed, whereas support D is only partly restrained and can be assumed to have a support moment corresponding to half the fixed end moment. The other two supports are freely supported. It is assumed that no moment is transferred from the column to the slab, i.e. the support moments on both sides of the column are the same.

The support moments are estimated to be

$$m_A \approx -\frac{8 \times 4.8^2}{12} = -15.36 \tag{8.1}$$

$$m_{xE} \approx -(\frac{8 \times 4.8^2}{12} + \frac{8 \times 5.2^2}{8})/2 = -21.20 \tag{8.2}$$

$$m_D \approx -\frac{1}{2} \times \frac{8 \times 5.6^2}{12} = -10.45 \tag{8.3}$$

146

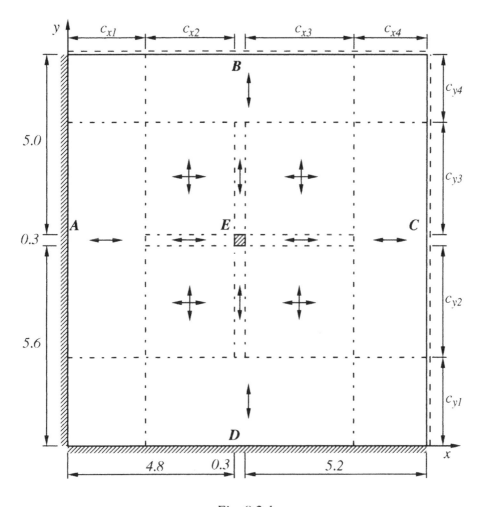

Fig. 8.2.1

$$m_{y\,E} \approx -\left(\frac{8 \times 5.6^2}{10} + \frac{8 \times 5.0^2}{8}\right)/2 = -25.04 \qquad (8.4)$$

If we wish, we may choose support moments which are somewhat larger or smaller, but in this case we just choose the calculated values. We can now calculate the c-values from Eq. (2.34):

$$c_{x2} = \frac{4.8}{2} + \frac{21.20 - 15.36}{8 \times 4.8} = 2.55 \tag{8.5}$$

$$c_{x3} = \frac{5.2}{2} + \frac{21.20}{8 \times 5.2} = 3.11 \tag{8.6}$$

$$c_{y2} = \frac{5.6}{2} + \frac{25.04 - 10.45}{8 \times 5.6} = 3.13 \tag{8.7}$$

$$c_{y3} = \frac{5.0}{2} + \frac{25.04}{8 \times 5.0} = 3.13 \tag{8.8}$$

We can now calculate the average span moments from Eq. (2.35):

$$m_{AE} = \frac{8 \times 2.55^2}{2} - 21.20 = 4.81 \tag{8.9}$$

$$m_{EC} = \frac{8 \times 3.11^2}{2} - 21.20 = 17.49 \tag{8.10}$$

$$m_{DE} = \frac{8 \times 3.13^2}{2} - 25.04 = 14.15 \tag{8.11}$$

$$m_{EB} = \frac{8 \times 3.13^2}{2} - 25.04 = 14.15 \tag{8.12}$$

Provided that we use moment distributions chosen from $s1$, $s2$ and $f1$ (corresponding to $\beta=0.5$) in Fig. 8.1.4 we may choose the distribution of support moments arbitrarily except with regard to span AE, where the ratio between support and span moment is $-21.20/4.81 = -4.41$. If we choose distribution $s1$ for the support moment we find $\alpha = 4.81/(4.81+21.20) = 0.18$, which is too low according to Eq. (2.25). We have to take at least $-0.25(4.81+21.20) + 4.81 = -1.69$ as support moment in the middle strip.

As the span moment in AE comes out so low it might have been better to choose a little lower value of the support moment at A, e.g. -12. The difference in behaviour of the slab with such a change is probably quite insignificant.

The support reaction at the column is $8(2.55+0.3+3.11)(3.13+0.3+3.13) = 313$ kN.

8.2.2 More than one interior column

In a regular flat slab all columns should be situated at the crossing points of lines parallel to the edges. The design procedure will be the same regardless of the number of columns. It will be demonstrated on a slab with only two interior columns.

Example 8.2

The slab in Fig. 8.2.2 carries a load of 11 kN/m². Supports A and B are fixed, C and D freely supported. The columns have strong column capitals with a diameter of 2.0 m.

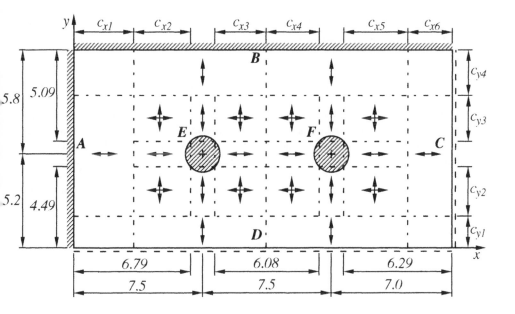

Fig. 8.2.2

The spans are determined as if the interior supports consist of rectangles inscribed into the circular column capitals. A a square is usually chosen, and this will be done here. A slightly better choice would have been to use a rectangle, for example, with length 1.6 m in the x-direction and width 1.2 m in the y-direction. This would have decreased the reinforcement in the x-direction by a few percent and increased it in the y-direction by a few percent and given a total decrease in reinforcement of between 1 and 2 percent, which is insignificant.

With the inscribed square as the support we get the spans given in the figure. With these spans we can calculate the approximate magnitudes of the support moments and choose the design values of these moments:

$$m_A \approx -\frac{11 \times 6.79^2}{12} = -42.26 \qquad \text{choose} \ -43.0 \qquad (8.13)$$

149

$$m_{xE} \approx -\frac{11 \times 6.79^2}{12} + \frac{11 \times 6.08^2}{12} / 2 = -38.07 \qquad \text{choose } -35.0 \qquad (8.14)$$

$$m_{xF} \approx -\left(\frac{11 \times 6.08^2}{12} + \frac{11 \times 6.29^2}{8}\right) / 2 = -44.14 \qquad \text{choose } -40.0 \qquad (8.15)$$

$$m_{yE} = m_{yF} \approx -\left(\frac{11 \times 4.49^2}{8} + \frac{11 \times 5.09^2}{12}\right) / 2 = -25.73 \qquad \text{choose } -24.0 \qquad (8.16)$$

$$m_B \approx -\frac{11 \times 5.09^2}{12} = -23.75 \qquad \text{take } -23.75 \qquad (8.17)$$

The values of support moments at the columns have been chosen somewhat lower than the calculated values because of the large column capitals. The value of m_A has been chosen a little higher than the calculated value because the moment at E is a little lower than the value for a fixed support. The choice of values may be discussed, but small variations will not influence the behaviour of the slab.

The c-values are calculated from Eq. (2.34) and the corresponding span moments from Eq. (2.35):

$$c_{x1} = \frac{6.79}{2} + \frac{43.0 - 35.0}{11 \times 6.79} = 3.50 \qquad (8.18)$$

$$c_{x3} = \frac{6.08}{2} + \frac{35.0 - 40.0}{11 \times 6.08} = 2.97 \qquad (8.19)$$

$$c_{x5} = \frac{6.29}{2} + \frac{40.0}{11 \times 6.29} = 3.72 \qquad (8.20)$$

$$c_{y1} = \frac{4.49}{2} - \frac{24.0}{11 \times 4.49} = 1.76 \qquad (8.21)$$

$$c_{y3} = \frac{5.09}{2} + \frac{24.0 - 23.75}{11 \times 5.09} = 2.55 \qquad (8.22)$$

$$m_{AE} = \frac{11 \times 3.50^2}{2} - 43.0 = 24.4 \qquad (8.23)$$

$$m_{EF} = \frac{11 \times 2.97^2}{2} - 35.0 = 13.5 \qquad (8.24)$$

$$m_{FC} = \frac{11 \times 3.72^2}{2} - 40.0 = 36.1 \tag{8.25}$$

$$m_{DE} = m_{DF} = \frac{11 \times 1.76^2}{2} = 17.0 \tag{8.26}$$

$$m_{EB} = m_{FB} = \frac{11 \times 2.55^2}{2} - 24.0 = 11.8 \tag{8.27}$$

In this case the ratios between support and span moments are such that the distribution of design moments can be chosen rather freely. As the one-way elements between the corner-supported elements are rather wide it is acceptable to use a width of the column strip which is wider than half the width of the individual strip. As the spans in the y-direction are much smaller than the spans in the x-direction the relative width of the column strip in the y-direction may be chosen smaller, e.g. with the same width as in the x-direction.

8.3 Exterior column supports

8.3.1 General

In this section the emphasis is on the design of a slab where the edge of the slab is supported by columns. In order to simplify the examples, slabs are considered without any interior column, as it is not difficult to combine the design procedure demonstrated below with the procedure above.

Where it is assumed that no moment transfer takes place between the exterior column and the slab it must still be remembered that most of the support reaction is assumed to act at the column face. This eccentricity causes a moment in the column.

8.3.2 Column support at one edge

Example 8.3

The slab in Fig. 8.3.1 carries a load of 9 kN/m^2. All supported edges are freely supported. It is also assumed that there is no moment transfer to the column.

The only support moment is the moment m_{xB} above the column. The value is estimated from

$$m_{xB} \approx -(\frac{9 \times 4.7^2}{8} + \frac{9 \times 5.2^2}{8})/2 = 27.64 \tag{8.28}$$

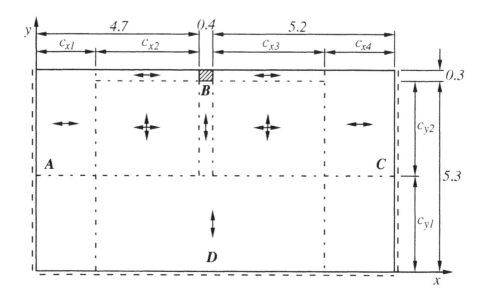

Fig. 8.3.1

We may choose $m_{xB} = -27.0$. From Eqs. (2.34) and (2.35) we now calculate the c-values and the span moments:

$$c_{x1} = \frac{4.7}{2} - \frac{27.0}{9 \times 4.7} = 1.71 \tag{8.29}$$

$$c_{x4} = \frac{5.2}{2} - \frac{27.0}{9 \times 5.2} = 2.02 \tag{8.30}$$

$$c_{y1} = \frac{5.3}{2} = 2.65 \tag{8.31}$$

$$m_{AB} = \frac{9 \times 1.71^2}{2} = 13.16 \tag{8.32}$$

$$m_{BC} = \frac{9 \times 2.02^2}{2} = 18.36 \tag{8.33}$$

$$m_{DB} = \frac{9 \times 2.65^2}{2} = 31.60 \tag{8.34}$$

The ratio between support and span moments permits a rather arbitrary distribution of reinforcement in the x-direction of the types shown in Fig. 8.1.4. It is acceptable to concentrate some reinforcement to the vicinity of the free edge (within the limits given by Eq. (2.24)), particularly some span reinforcement. On the other hand the one-way element theoretically requires an evenly distributed design moment, so the span moment should not be too unevenly distributed.

In the y-direction there is no support moment, which means that the condition in Eq. (2.24) has to be fulfilled by the span reinforcement alone. Thus the design span reinforcement in the middle strip may not be larger than $0.7m_{DE}$ and the remaining part must be concentrated in the column strip. This of course means that the reinforcement is not evenly distributed in the one-way element at edge D, but this is of no practical importance.

There should be some bars along the free edge and some stirrups in order to avoid torsional cracks.

An example of a suitable distribution of the moments in the left-hand half of the slab can be seen in Fig. 8.3.3.

Example 8.4

The slab and load are the same as in the preceding example, but with a moment transfer to the column. It is assumed that the average moment is –8.0 kNm/m at the section through the inner edge of the column, which gives a total moment of –8.0(2.99+0.4+3.18) = –52.6 kNm. How this moment is calculated or estimated is not discussed here.

The analysis is exactly as above regarding the x-direction. For the y-direction we get:

$$c_{y1} = \frac{5.3}{2} - \frac{8.0}{9 \times 5.3} = 2.48 \tag{8.35}$$

$$m_{DB} = \frac{9 \times 2.48^2}{2} = 27.68 \tag{8.36}$$

It is recommended that the top reinforcement corresponding to the moment in the column is concentrated in a small width. It has to be anchored in such a way that the continuity is secured. As the support moment is rather small the condition in Eq. (2.24) is not fulfilled if the span reinforcement is evenly distributed. The span design moment for the middle strip may not be larger than 0.7(27.68+8.0) = 24.98 and the remaining part has to be concentrated in a column strip of a suitable width.

There should be some bars along the free edge and some stirrups in order to avoid torsional cracks.

8.3.3 Column support at a corner

Example 8.5

Fig. 8.3.2 shows a slab which is similar to the slab in the preceding two examples except that the supporting wall C is exchanged for a column at the outer corner. If we assume that the slab is freely supported at the side of the columns the c-values and the average moments are exactly the same as in Example 8.3. The only difference is the distribution of design moments.

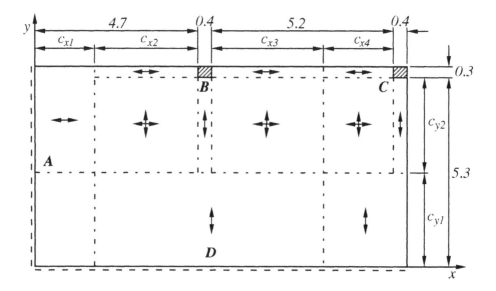

Fig. 8.3.2

As the corner-supported element at the corner column has no support moment the span reinforcement has to be distributed in such a way that condition (2.24) is fulfilled. This means that not more than $0.7m_{BC}$ may be taken in the middle strip. In the x-direction it is, however, satisfactory to take less than this in the middle strip and more in the column strip, i.e. along the free edge, and also to make the column strip smaller than half the strip width. In this way some reinforcement is concentrated along the free edge, which is to be recommended, particularly in this case where the strip spans between two columns.

As the slab is assumed to be simply supported at the corner column there is no design top reinforcement in that region. There will, however, be certain negative moments under service conditions, which may cause cracks, mainly in the direction of the bisector to the corner. It is advisable to have some top reinforcement at right angles to that direction.

In the y-direction the distribution of span moments must also be chosen so as to fulfil condition (2.24), i.e. with not more than $0.7m_{DB}$ in the middle strip. In this case the cooperating element at edge D is a one-way element, which means that it is not acceptable to have too uneven a distribution of span moment. For this direction the moment in the middle strip may thus be chosen to be $0.7m_{DB}$ or somewhat smaller.

Fig. 8.3.3 shows a possible distribution of design moments.

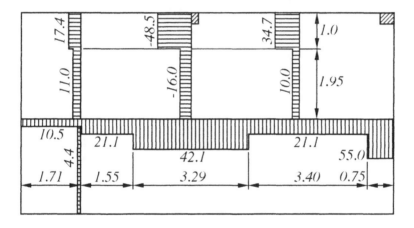

Fig. 8.3.3

In most practical cases it is recommended to take into account the moment transfer to the corner column as well as to the edge column. The influence of these moments on the design is demonstrated in Example 8.4.

8.4 Slab cantilevering outside columns

Where the slab cantilevers outside a column we get corner-supported elements without any positive moment corresponding to the reinforcement in the direction of the cantilever. Then the support design moment has to have a distribution which satisfies condition (2.24).

Example 8.6

The slab in Fig. 8.4.1 carries a load of 14 kN/m² and is supported on column capitals with a diameter of 1.5 m. The support is taken as the inscribed square, which has a side length of 1.06 m. This gives the span lengths shown in the figure. No moment transfer is assumed to take place between the slab and the columns.

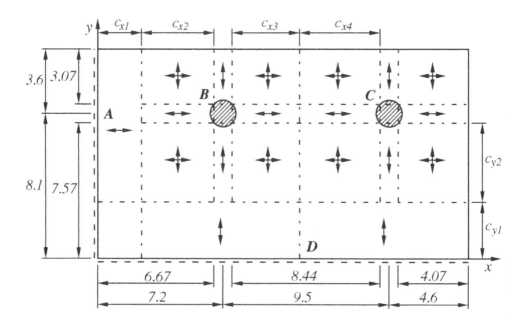

Fig. 8.4.1

The average moments caused by the cantilevers are statically determinate. We thus get

$$m_{x\,C} = -\frac{14 \times 4.07^2}{2} = -116.0 \qquad (8.37)$$

$$m_{yB} = m_{yC} = -\frac{14 \times 3.07^2}{2} = -66.0 \qquad (8.38)$$

The support moment m_{xB} is statically indeterminate and can be estimated from

$$m_{xB} \approx - (\frac{14 \times 6.67^2}{8} + \frac{14 \times 8.44^2}{12}) / 2 = -80.5 \qquad \text{choose } -75.0 \qquad (8.39)$$

The reason why m_{xB} has been chosen lower than the approximate value is partly the large column capital and partly the fact that m_{xC} is larger than the moment corresponding to a fixed support for span BC. It might even have been reduced further.

We can now calculate the c-values and the corresponding span moments from Eqs (2.34) and (2.35):

$$c_{x1} = \frac{6.67}{2} - \frac{75.0}{14 \times 6.67} = 2.53 \qquad (8.40)$$

$$c_{x3} = \frac{8.44}{2} + \frac{75.0-116.0}{14 \times 8.44} = 3.87 \qquad (8.41)$$

$$c_{y1} = \frac{7.57}{2} - \frac{66.0}{14 \times 7.57} = 3.16 \qquad (8.42)$$

$$m_{AB} = \frac{14 \times 2.53^2}{2} = 44.8 \qquad (8.43)$$

$$m_{BC} = \frac{14 \times 3.87^2}{2} - 75.0 = 29.8 \qquad (8.44)$$

$$m_{DB} = m_{DC} = \frac{14 \times 3.16^2}{2} = 69.9 \qquad (8.45)$$

Fig. 8.4.2 shows a possible distribution of design moments for the reinforcement in the x-direction. It fulfils condition (2.24) and is suitable for crack limitation. It corresponds to $\beta=0.5$. If, for example, we check the α-value to the left of column B we find

$$\alpha = \frac{44.8 + 37.5}{44.8 + 75.0} = 0.69 < 0.7 \qquad (8.46)$$

Thus we are close to the limit, and we could not have increased the support moment at the edge without decreasing the corresponding span moment to an equal extent.

In this case it is not recommended that all the support moment should be concentrated in the column strip, as this would leave the edge without top reinforcement.

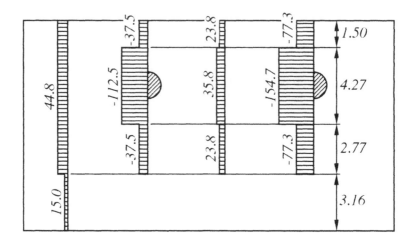

Fig. 8.4.2

8.5 Oblong panels and corner-supported elements

The advanced strip method always results in a design which is safe, which means that the slab has adequate safety against bending failure. For the design it is also important that the slab functions well under service conditions and that the reinforcement economy is good. For most slabs met with in practice this can be expected to be the case if the rules above regarding moment ratios and lateral moment distributions are followed.

If, however, the panels are very elongated the advanced strip method may lead to a moment distribution which is too different from the expected distribution in the service state. This might lead to unacceptable cracks in places where the moments have been underestimated. In such cases it may be better to assume that the slab is composed of ordinary rectangular slabs supported on support bands.

Fig. 8.5.1 demonstrates an example. Fig. 8.5.1 a) shows elements for a design based on the advanced strip method. The panels studied in this approach, e.g. *ABCD*, are very elongated, as well as some of the corner-supported elements.

Fig. 8.5.1 b) shows elements for a design of the same slab assuming rectangular slabs supported on support bands. Thus the rectangular slab *ABEF* is assumed to be supported along all sides and continuous at *ADF*. The supports along *AF* and *BE* consist of support

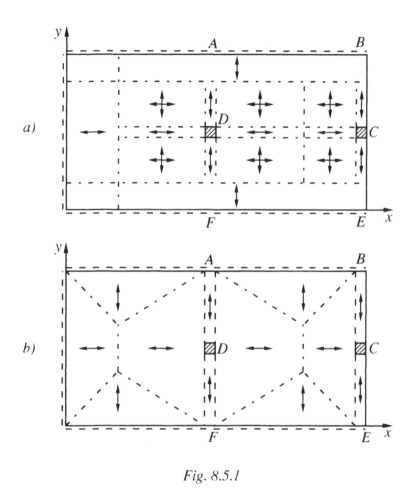

Fig. 8.5.1

bands which are continuous over columns D and C respectively. The design follows the principles demonstrated in Chapter 4.

If the design moments obtained from the two approaches are compared it is found that a) gives larger moments m_x and m_{ys} but smaller moments m_{yf} than b). The risk with the application of the advanced strip method thus lies in an underestimation of m_{yf} leading to possible cracking in the underside of the slab. This risk increases the more the shapes of the panels and the corner-supported elements deviate from that of a square. It is not possible to give any general rule for when the risk has to be taken into account. In practice this will probably seldom be the case, as the column supports are normally arranged so that the spans in the

159

two directions are not too unequal. A practical case may be where a free edge of a rectangular slab is supported on one or more columns. In this case the approach according to b) is normally to be preferred, although the approach according to a) may formally be applied.

In cases where there is a doubt whether approach a) gives an acceptable result it is recommended that both approaches are used and the results compared. If the differences are great it is advisable to use approach b) or to take some kind of weighted average between the results.

It must be remembered that the choice of approach is unimportant with regard to the ultimate limit state and that it has only to do with crack control.

CHAPTER 9

Regular flat slabs with non-uniform loads

9.1 Introduction

Where the load on a flat slab is not uniform the advanced strip method in its basic form can only be used if the load is uniform in one of the main directions. Acceptable approximate solutions with the method can, however, be found.

The reason why the advanced strip method in its basic form is not generally applicable for non-uniform loads is that the lines of zero shear force are not continuous, so the widths of the strips will vary.

9.2 Uniform loads in one direction

If the load on a flat slab is uniform in one of the main directions the advanced strip method can be applied, as the lines of zero shear force are straight and continuous. The only differences from the analysis of flat slabs with a uniform load are that the load varies in one direction and the moments in parallel strips in the other direction differ, i.e. they have different loads. The moments in these strips are chosen so that the maximum span moments occur along the same line.

Example 9.1

The load on the slab in Fig. 9.2.1 is 9 kN/m^2 between lines a and e, and 17 kN/m^2 between lines e and i. The edge along line l is fixed, whereas the other edges are freely supported. All columns are 0.4m×0.4 m.

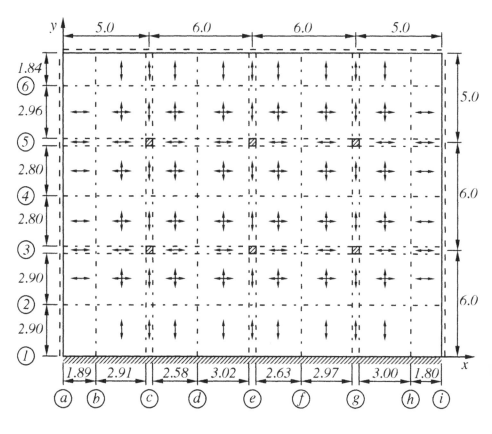

Fig. 9.2.1

The strips in the x-direction have different loads on the spans $a–e$ and $e–i$. An approximate elastic analysis shows that suitable values of support moments are:

$m_c = -22.0, m_e = -33.0, m_g = -49.0.$

Based on these support moments the c-values given in the figure have been calculated by means of Eq. (2.34). Eq. (2.35) gives the span moments:

$m_b = 16.1$, $m_d = 8.0$, $m_f = 25.9$, $m_h = 27.5$.

The strips in the y-direction to the right of line e have a uniform load of 17 kN/m². An approximate elastic analysis shows that the support moments can be chosen to be equal:

$m_1 = m_3 = m_5 = -46.0$.

The corresponding c-values are given in the figure. The span moments are:

$m_2 = 25.5$, $m_4 = 20.6$, $m_6 = 28.7$.

The strips to the left of line e have a uniform load of 9 kN/m². The moments are simply taken as 9/17 of those to the right of line e:

$m_1 = m_3 = m_5 = -24.4$, $m_2 = 13.5$, $m_4 = 10.9$, $m_6 = 15.2$.

The distributions of design moments follow the recommendations in Section 8.1.5. One special question in this case is the support moments for the reinforcement in the y-direction above the columns in line e. The most correct distribution is to follow the calculated moments strictly and use different values to the left and right of the column centre. This is what is recommended in the first place.

Suitable distributions of design moments along lines *1*, *2* and *3* are shown in Fig. 9.2.2. The distributions correspond to $\beta = 0.5$, which means that the widths of the column strips are equal to half the total strip widths. Other distributions are possible, depending on the need for crack control.

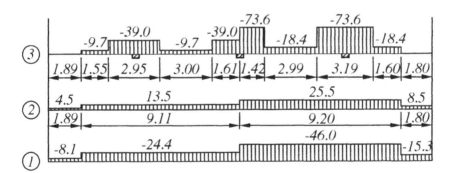

Fig. 9.2.2

A check should in principle always be made that the rules for lateral moment distributions in Section 2.5.2 are followed. With the normally recommended distribution of support and span moments such checks need, however, only be made if the ratio between the numerical values of the support and span moments is above about 3 or below about 1. For this slab such a check only has to be made for the moments m_x in element d–e, where this ratio is 33.0/8.0 = 4.1. If we choose an even distribution of span moment and concentrate all the support moment in the column strip we get $\alpha = 8.0/(33.0+8.0) = 0.20$. This is not an acceptable value according to Eq. (2.24). We have to take some of the support moment in the middle strip. We may, for example, take $0.4m_s$ in the middle strip, as has been done in Fig. 9.2.2. Then we get $\alpha = (8.0+0.4\times33.0)/(8.0+33.0) = 0.52$, which is satisfactory.

From the point of view of construction it may be advantageous to use only one design value for the support reinforcement over each column. Instead of the values -39.0 and -73.6 a weighted mean value might be used over the whole width for the column strip in line e. This weighted mean value is

$$-\frac{39.0 \times 1.61 + 73.6 \times 1.42}{1.61 + 1.42} = -55.2 \qquad (9.1)$$

The safety and behaviour of the slab would probably not be much influenced by choosing the average value instead of the two values in Fig. 9.2.2, but in order to avoid any risks it is recommended that the two values given in the figure are used.

9.3 Different loads on panels

The advanced strip method can be applied to the situation where different loads are applied to panels in an approximate but safe way by not keeping the regularity of the net of lines of zero shear force in the spans and accepting that the lines may change position where they pass from one strip to another. This means that the strips change width where they pass a support. The strips on both sides of the support do not fit together. Even if they have the same average moments there is not perfect equilibrium over the support as the total moments are different.

The simplest way of treating this problem is to reinforce for the largest total moment over the support, i.e. to reinforce for the support moment on the width of the widest strip. In this way the design is on the safe side. In practice the difference in width is seldom great, which means that the economical consequence of this simple approach is unimportant.

Example 9.2

Fig. 9.3.1 shows a slab with different loads on different areas. The slab rests on large column heads and is freely supported along all edges.

The strips which are used for the design are shown in Fig. 9.3.2, where all the c-values which have resulted from the analysis are given.

In the x-direction the strip between lines *2* and *3* has one span with load 8 kN/m^2 and two spans with 18 kN/m^2. An approximate elastic analysis of the strip shows that suitable support moments are $m_c = -45$ and $m_e = -65$. With these support moments we get the c-values given in the figure from Eq. (2.34) and span moments $m_b = 9.2$, $m_d = 37.4$, $m_f = 32.7$ from Eq. (2.35).

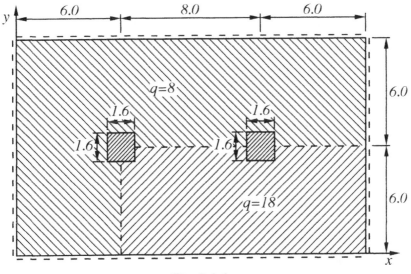

Fig. 9.3.1

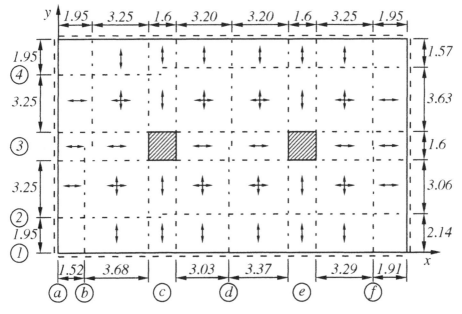

Fig. 9.3.2

In the same way, for the strip between lines *3* and *4* we find the moments $m_c = m_e = -27$, $m_b = m_f = 15.2$, $m_d = 14.0$.

For the strip in the y-direction between lines *b* and *c* we find $m_3 = -27$, $m_2 = m_4 = 15.2$.

For the strip between lines *c* and *f* we find $m_3 = -43$, $m_2 = 41.2$, $m_4 = 9.8$.

Figs 9.3.3 and 9.3.4 show proposed distributions of design moments for the reinforcement in the x- and y-directions, respectively.

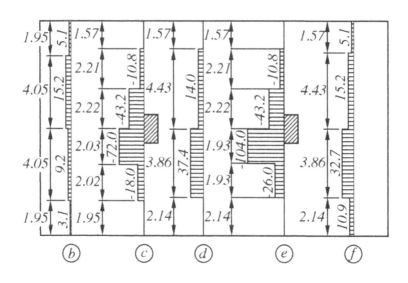

Fig. 9.3.3

The span moments have been chosen to be uniform within each strip and to have the values above. For the distribution of support moments the column strips and middle strips have been assumed to have equal widths, corresponding to $\beta = 0.5$. The support moments in the middle strips have been taken as 0.4 times, and in the column strips 1.6 times, the average moment in the strip to which they belong. The distribution of support moments thus corresponds to case *s2* in Fig. 8.1.4.

Just as in the previous example the support moments in the column strips might be evenly distributed and equal to a weighted average, but in the first place it is recommended that the distributions shown in the figures are used.

A check against the rules in Section 2.5.2 has in this case to be performed for the moments $m_{xc} = -45$ and $m_{xb} = 9.2$, where the ratio between the numerical values of the support and span moments is greater than 3. With the distribution in Fig. 9.3.3 we have $\alpha = (9.2+18.0)/(9.2+45) = 0.50$, which is acceptable according to Eq. (2.24).

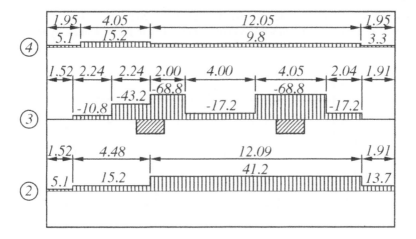

Fig. 9.3.4

9.4 Concentrated loads

Where more or less concentrated loads are acting on a flat slab they are taken into account in the determination of the average support and span moments in the strips just like uniform loads.

If the concentrated load is small compared to the total uniform load on the same panel, say less then 10%, the normal lateral distribution of design moments may be used. It may be somewhat better for the behaviour under service conditions to make some redistribution of moment to the part of the panel where the concentrated load is acting, and thus to a column strip if the concentrated load is acting mainly within that strip.

If the concentrated load is not small compared to the total uniform load on the panel, the analysis is best made separately for the uniform load and for the concentrated load and the moments are then added. This leads to some overestimation of the span moments, as the maximum moments from the two loading cases do not appear in the same sections.

The procedure for determining the design moments caused by a concentrated load is best illustrated by means of an example.

<u>Example 9.3</u>

Fig. 9.4.1 shows an interior panel within a flat slab with a concentrated load $F = 120$ kN acting over an area 1.2m×1.2 m with its centre at (3.0/1.0) in the coordinate system shown.

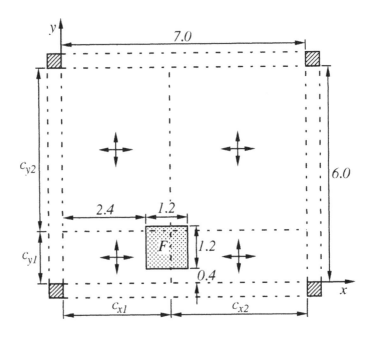

Fig. 9.4.1

We start by determining the total moments caused by the concentrated load. The support moments may be taken as about half the moments according to the theory of elasticity for a beam with the same load and fixed ends. We may choose the values

$M_{xs1} = -59$, $M_{xs2} = -44$, $M_{ys1} = -38$, $M_{ys2} = -8$ kNm,
where index *1* indicates the ends on the coordinate axes.
Based on these values we find the support reactions

$$R_{x1} = (120 \times 4.0 + 59 - 44)/7.0 = 70.7 \text{ kN} \tag{9.2}$$

$$R_{y1} = (120 \times 5.0 + 38 - 8)/6.0 = 105.0 \text{ kN} \tag{9.3}$$

The load per m length in the strips in both directions is equal to $120/1.2 = 100$ kN/m. The positions of the lines of zero shear force are determined from

$$c_{x1} = 70.7/100 + 2.4 = 3.11 \tag{9.4}$$

$$c_{y1} = 105.0/100 + 0.4 = 1.45 \tag{9.5}$$

The span moments are

$$M_{xf} = 70.7(2.4 + 3.11)/2 - 59 = 135.8 \tag{9.6}$$

$$M_{yf} = 105.0(0.4 + 1.45)/2 - 38 = 59.1 \tag{9.7}$$

The lines of zero shear force divides the load between the different elements and different parts of the strip. Thus $1.05/1.20$ of the load is carried on the strip with width c_{y1}. The corresponding parts of the moments are distributed on the widths of the strips in question. We get the following average design moments:
In the strip with width c_{y1}:

$$m = \frac{1.05}{1.20 \times 1.45} \times M = 0.6034\,M \tag{9.8}$$

$$m_{xs1} = -0.6034 \times 59 = -35.60 \tag{9.9}$$

$$m_{xs2} = -0.6034 \times 44 = -26.55 \tag{9.10}$$

$$m_{xf} = 0.6034 \times 135.8 = 81.95 \tag{9.11}$$

In the same way we find:
In the strip with width c_{y2}: $m_{xs1} = -1.62, m_{xs2} = -1.21, m_{xf} = 3.73$
In the strip with width c_{x1}: $m_{ys1} = -7.23, m_{ys2} = -1.52, m_{yf} = 11.24$
In the strip with width c_{x2}: $m_{ys1} = -3.99, m_{ys2} = -0.84, m_{yf} = 6.20$
It can be seen that the dominant moments are those in the strip with width c_{y1}. These moments are rather high and concentrated in a narrow band. It may be better to distribute them in a wider band. This can be done through the introduction of distribution reinforcement in the y-direction, applying the principle demonstrated in Section 2.6.1. Let us distribute the load over a distance of $b=3.0$ m in the y-direction with a bandwidth $a=2.0$ m. Eq. (2.33) gives a design moment for the distribution reinforcement

$$m_{yf} = \frac{120\,(3.0 - 1.2)}{8 \times 2.0} = 13.5 \tag{9.12}$$

The load is now acting on the slab over an area of $1.2\text{m} \times 3.0$ m. One part of this area is outside the studied corner-supported elements and acts on the parallel one-way element in

169

the x-direction (or slightly on the other side of that element). The part of the load acting on the corner-supported element is $120 \times 2.5/3.0 = 100$ kN, and it has its centre at $(3.0/1.25)$.

The total moments M_x (including the one-way element) and the value of c_{x1} are unchanged. It proves satisfactory to choose $M_{ys1} = -34$, $M_{ys2} = -5$. With these moments we find $R_{y1} = 84.0$, $c_{y1} = 2.10$, $M_{yf} = 54.2$.

As only $2.5/3.0$ of the width in the y-direction is within the corner-supported element only $2.5/3.0$ of the moments M_x belong to these elements. For the strip with width c_{y1} we thus get the ratio between m and M:

$$m = \frac{2.5}{3.0} \times \frac{2.1}{2.5 \times 2.1} \times M = 0.3333 \, M \qquad (9.13)$$

In this way we get the following average design moments:
In the strip with width c_{y1}: $m_{xs1} = -19.67$, $m_{xs2} = -14.67$, $m_{xf} = 45.27$. The parallel one-way element has the same moments.
In the strip with width c_{y2}: $m_{xs1} = -2.02$, $m_{xs2} = -1.50$, $m_{xf} = 4.64$.
In the strip with width c_{x1}: $m_{ys1} = -6.47$, $m_{ys2} = -0.95$, $m_{yf} = 10.31$.
In the strip with width c_{x2}: $m_{ys1} = -3.57$, $m_{ys2} = -0.53$, $m_{yf} = 5.69$.

Now the largest moments per unit width are reduced to about 55% of the earlier values at the expense of some distribution reinforcement.

The distribution of m_f- m_s for a point load will be uniform according to Section 2.6.2. By integration over the loaded area a uniform distribution for a point load corresponds to a triangular distribution on the width of a uniform load and a constant value on the remaining part, according to Fig. 9.4.2. This distribution may be transformed to stepped constant values. It is suitable to distribute m_s with a certain concentration towards the supported corner and m_f somewhat towards the load. For the final distribution of design moments some further redistribution may be accepted in order to simplify.

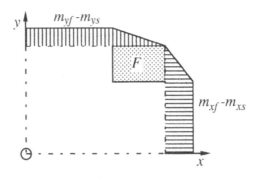

Fig. 9.4.2

Fig. 9.4.3 shows a possible distribution of design moments based on the second solution. These moments are to be added to the moments caused by the uniform load on the slab.The reinforcement should not, in principle, be curtailed within a corner-supported element in this case.

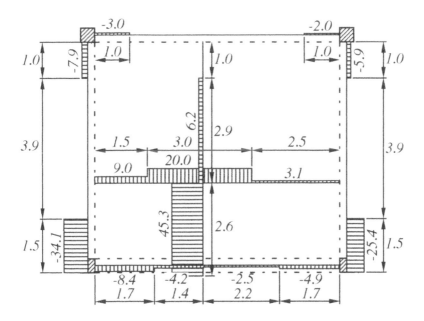

Fig. 9.4.3

The support moments have to be taken care of in the adjacent panels. This will cause a change in the position of the lines of zero shear force in the spans. The span moments may thus be reduced. The increase in length of reinforcing bars should be taken into account.

171

CHAPTER 10

Irregular flat slabs

10.1 General

An irregular flat slab may have columns placed quite arbitrarily and edges at any angle to each other. It is of course not possible to find simple general methods for the design of all types of such slabs, automatically leading to suitable results from all points of view, particularly regarding reinforcement economy and behaviour under service conditions.

Whatever approach is used within the framework of the strip method, it is mostly necessary to use an iteration process in order to get a result with a satisfactory distribution of reinforcement. Estimating whether the reinforcement is satisfactory is mainly based on a comparison between support and span moments and on some feeling for the behaviour of a slab.

One possible approach might be the use of support bands between the columns, supporting strips which in their turn carry the loads. This approach is generally complicated because there are so many choices to be made, and therefore so many iterations. A major problem is that the support bands which meet over a column have different directions of support moments, which have to be taken by straight reinforcement bars. This approach is therefore not generally satisfactory.

An alternative approach will be recommended here which is based on the restricting condition that *most reinforcing bars are placed in two orthogonal directions*. This restriction is valid for span as well as support reinforcement. An exception is reinforcement along free edges, which is best placed along the edge. Another exception is that support reinforcement at a support where the slab is fixed or continuous is best arranged at right angles to the support.

The main advantage of this approach is that it gives the possibility of formulating general rules for the design, which can be carried through in a systematic way. Another advantage is that it leads to a simple reinforcement arrangement.

The main disadvantage is that the reinforcement directions may deviate considerably from the directions of the principal moments. Where this is the case the reinforcement is less efficient, which means that the crack control is not as good as when the reinforcement directions coincide with the principal moment directions. It may also mean a poorer reinforcement economy. On the other hand, reinforcement which follows the directions of the principal moments better is more complicated and may be more expensive.

Over a column the moments often tend towards a polar symmetry. Where this is the case the reinforcement directions are not important, provided that there are two orthogonal reinforcement directions.

From a practical point of view the advantages of the approach are probably greater than the disadvantages.

The approach is not generally applicable to slabs with free edges. This will be discussed in connection with some examples.

10.2 Design procedure

The proposed approach, which is based on the condition that all reinforcing bars within the interior of the slab are placed in two directions at right angles, parallel to the coordinate axes, normally contains the following steps:

1. Determine suitable directions of the coordinate axes, corresponding to the reinforcement directions. One axis may, for instance, be chosen parallel to an edge of the slab or to an important direction between column centres.

2. Determine theoretical column profiles. The theoretical column profile for each column is an inscribed rectangle with sides parallel to the coordinate axes.

3. Determine the lines of zero shear force where the span moments have their maxima, the *span lines*, in the following way. First, place lines of zero shear force between the edges and the columns nearest the edges. These lines are drawn at about half the distance between the column and the edge if the edge is fixed, and at about 0.625 of this distance from the columns if the edge is freely supported. If the edge is only partly restrained, use intermediate values. Then draw lines (these are not lines of zero shear force) between adjacent column centres. Starting from the centres of these lines and running at right angles to them, draw lines of zero shear force. Draw them in both directions until they intersect the corresponding lines from other pairs of columns or other lines of zero shear force. In cases where two or more intersections are close to each other adjust the lines so that they have a common point of intersection.

4. Draw the lines of zero shear force where the support moments have their maxima. These *support lines* are a continuation of the sides of the theoretical column profiles and are drawn in the directions of the coordinate axes until they meet the span lines. There are four support lines around each column (except for edge and corner columns), two in each direction. There is also a support line at each support on a wall, beam or support band.

5. Calculate the total moments (span minus support moment) with respect to each of the support lines at the columns. Each moment is caused by the load on the element formed by the support line in question and the nearest span lines.The span moments are best expressed as average moments (m) but the support moments at columns as total moments (M). Calculate also the corresponding average moments in the one-way elements supported on walls, beams or support bands. Where such a support is not parallel to a coordinate axis the span moment m with respect to the support has to be taken by design span moments m_x and m_y in the reinforcement directions. The following relation is valid:

$$m_x \sin^2 \varphi + m_y \cos^2 \varphi = m \qquad (10.1)$$

where φ is the angle between the support and the x-axis. Often it is suitable to assume $m_x = m_y = m$. It is also possible to increase one of the design moments and decrease the other according to the equation.

6. Choose support moments M_x and M_y at each column where the slab is continuous in the relevant direction. Values approximately equal to one-third (between 0.30 and 0.37) of the sum of the relevant total moments on both sides of the column will be suitable. These values are usually close enough to the support moment, according to the theory of elasticity. Choose (or calculate, if statically determined) support moments where the slab is supported by a wall, beam or support band. If the slab is fixed at such a support the support moment can be taken as approximately two-thirds (between 0.60 and 0.75) of the sum of the relevant total moments in the element.

7. Calculate the span moments from the values according to steps 5 and 6. This will give two values of each span moment, one from each side of the span line. Often there will be a great difference between these values. Then the span line has to be moved in the direction which will decrease the difference. A suitable distance Δl to move the span line can be estimated by means of the following relation

$$\Delta l \approx \frac{(m_1 - m_2) l}{8 (m_1 + m_2)} \qquad (10.2)$$

where m_1 and m_2 are the two span moments and l is the relevant span. Then go back to step 4 and make the analysis with the new positions of span and support lines. When the difference between the span moments at all span lines is small the result is accepted. The span moment might be taken as the average moment from the two values. This average moment will probably give an acceptable level of safety, but in order to be on the safe side a higher value may be preferred. To use the higher of the two values is always on the safe side, but may be unnecessarily conservative. A compromise is to add three-quarters of the higher value and one-quarter of the lower value. *This compromise is recommended and will be used in the examples.* An acceptable difference between the higher and lower value will be about 0.4 times the higher value in situations where this compromise is used. Where the span moment is small compared to the support moments (e.g. a short span between long spans) larger relative differences may be accepted.

8. Based on the calculated support and span moments, the design moments for the reinforcement are determined. The design span moments may be taken directly as the calculated average moments or redistributed towards places where the curvature is estimated to be larger. The support moment is suitably distributed with 70 to 80% over the central half of the support line and the rest distributed ouside this part, see the examples. A distribution with the whole moment concentrated within the central part may be accepted only if the ratios between the numerical values of support and span moments are everywhere between about 1.5 and 2.5.

A design based on these rules is theoretically not as safe as the design of regular flat slabs according to the rules given in Chapters 8 and 9, as the limitations on moment distributions used there are only valid for rectangular elements. However, the design is probably quite safe, particularly if the choice of moment distributions is also based on some feeling for the behaviour of slabs. *An exception is slabs with free edges, which have to be handled with care.* This is discussed in some examples.

A hand-calculation according to these rules takes a rather long time, as many simple operations have to be performed. It should not be too difficult to write a computer program which more or less automatically carries through the calculations, which may, perhaps, include some interaction with the designer.

10.3 Edges straight and fully supported

Example 10.1

The slab in Fig. 10.3.1 has a uniform load of 12 kN/m^2. It has three freely supported and one fixed edge and is supported on three columns. The column capitals have diameters of 1.0 m. The coordinates for the corners and the column centres are given in the figure in the chosen coordinate system, which has been assumed to give suitable reinforcement directions, parallel to the coordinate axes.

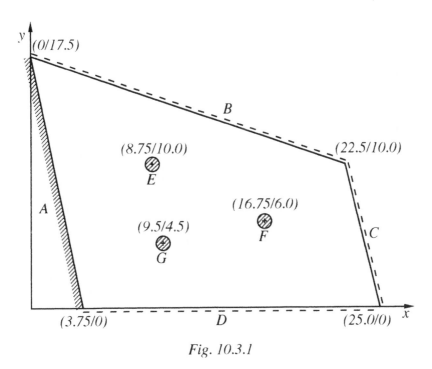

Fig. 10.3.1

The treatment of the slab according to the procedure above is illustrated in Fig. 10.3.2. The theoretical column support profiles are chosen as inscribed squares with edges parallel to the coordinate axes. These squares have sides 0.70 m. The types of element are indicated with arrows in the usual way.

First, the assumed span lines of zero shear force are drawn, starting with the lines *a-b-c-d* along the edges. Using the rules in point 3 above we find suitable coordinates for these points:

a (4.15/14.83), *b* (20.75/8.23), *c* (21.62/2.50), *d* (6.53/1.32)

Next, the lines between the centres of the columns are drawn (not shown in the figure) and span lines at right angles to these lines from their centres. These span lines (of zero shear force) are extended until they meet other span lines. In this way we find the following coordinates for the points of intersection:

e (13.83/1.89), *f* (5.57/6.78), *g* (12.61/7.72), *h* (14.15/10.85)

Now we have the complete pattern of span lines of zero shear force. Next, we draw the support lines of zero shear force along the sides of the theoretical column support profiles until they meet the span lines. The total number of points of intersection between support and span lines is 8 for each column and thus, in this case, 24 in all. These coordinate values

177

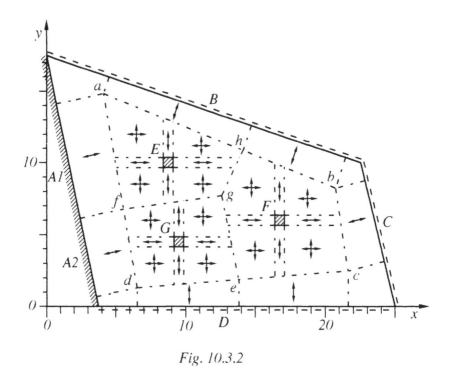

Fig. 10.3.2

are not all given here. Only some examples will be given in connection with moment calculations.

We also determine the distances at right angles from points a, b, c, d, e, f and h to the supports. These distances are shown as lines of zero shear force in Fig. 10.3.2.

Now we can write the equilibrium equations for the different parts (elements) of the slab. There are four such equations for each column and seven for the edge supports in this case, a total of 19 equations. Not all of these equations will be demonstrated in detail, but only a few typical examples. For the other elements only the resulting relations will be shown. All equilibrium equations are based on Eq. (2.7) with special cases (2.6) and (2.8).

The element between span line a-f and the support has $c = 3.50$ at a and $c = 3.20$ at f. This gives the following moment equation:

$$m_{AE} - m_{sA1} = \frac{12}{6}(3.50^2 + 3.50 \times 3.20 + 3.20^2) = 67.4 \qquad (10.3)$$

The support line to the left of column E has end coordinates (8.40/7.16) and (8.40/13.14) and thus a length of 5.98 m. The corresponding element has its corners at f and a. In writing the equilibrium equation it must be noted that the span moment has an active width equal to the length of the support line, as the reinforcement outside the ends of this line works in both directions and thus gives no contribution to the moment equation. In the analysis it is assumed that the span moment is uniformly distributed within each element. If another distribution is used for the design, this difference will normally make the design safer, as an uneven distribution of reinforcement is made in the direction where it is more efficient.

With the values above we get the following equation for the support line to the left of column E:

$$5.98\,m_{xAE} - M_{xE} = \frac{12}{6}[-2.83^2 \times 0.38 +$$

$$+(2.83^2 + 2.83 \times 4.25 + 4.25^2)8.05 - 4.25^2 \times 1.69] = 546 \qquad (10.4)$$

Continuing in the same way we find the following relations for the reinforcement in the x-direction:

$$5.60\,m_{xEF} - M_{xE} = 462 \qquad (10.5)$$

$$m_{AG} - m_{sA2} = 57.7 \qquad (10.6)$$

$$5.72\,m_{xAG} - M_{xG} = 327 \qquad (10.7)$$

$$5.77\,m_{xGF} - M_{xG} = 406 \qquad (10.8)$$

$$7.87\,m_{xEGF} - M_{xF} = 520 \qquad (10.9)$$

$$7.53\,m_{xFC} - M_{xF} = 629 \qquad (10.10)$$

$$m_{FC} = 34.7 \qquad (10.11)$$

Based on these relations we can calculate suitable support moments by means of the rule in point 6 in Section 10.2.

$M_{xE} = -(546+462)/3 = -336$, $M_{xG} = -(327+406)/3 = -244$, $M_{xF} = -(520+629)/3 = -383$, $m_{sA1} = -2\times67.4/3 = -44.9$, $m_{sA2} = -2\times57.7/3 = -38.5$.

Introducing these values into the equations above we get two values of each span moment. These two values are given below within parentheses as well as the weighted average according to the rule proposed in point 7 in Section 10.2. As the directions of edges A and C are not far from that of the y-axis it is, on the safe side, assumed that $m_y = 0$ in Eq. (10.1). Thus $0.956\,m_{xAE} = m_{AE}$, $0.941\,m_{xFC} = m_{FC}$ etc.

$m_{xAE} = (23.5/35.1) \Rightarrow 32.2$, $m_{xAG} = (20.1/14.5) \Rightarrow 18.7$, $m_{xEF} = (22.5/17.4) \Rightarrow 21.2$, $m_{xGF} = (28.1/17.4) \Rightarrow 25.4$, $m_{xFC} = (32.7/36.9) \Rightarrow 35.9$.

179

Performing the corresponding analysis for the y-direction gives the following results:

$$m_{yDG} = 15.6 \tag{10.12}$$

$$7.33\,m_{yDG} - M_{yG} = 288 \tag{10.13}$$

$$7.30 m_{yGE} - M_{yG} = 254 \tag{10.14}$$

$$8.50 m_{yGE} - M_{yE} = 262 \tag{10.15}$$

$$8.96 m_{yEB} - M_{yE} = 419 \tag{10.16}$$

$$m_{EB} = 14.1 \tag{10.17}$$

$$m_{yDF} = 29.1 \tag{10.18}$$

$$8.10 m_{yDF} - M_{yF} = 572 \tag{10.19}$$

$$8.14 m_{yFB} - M_{yF} = 514 \tag{10.20}$$

$$m_{FB} = 24.8 \tag{10.21}$$

$M_{yG} = -(288+254)/3 = -181$, $M_{yE} = -(262+419)/3 = -227$, $M_{yF} = -(572+514)/3 = -362$, $m_{yDG} = (15.6/14.6) \Rightarrow 15.4$, $m_{yGE} = (10.0/4.1) \Rightarrow 8.5$, $m_{yEB} = (21.4/15.7) \Rightarrow 20.0$, $m_{yDF} = (29.1/25.9) \Rightarrow 28.3$, $m_{yFB} = (18.7/27.6) \Rightarrow 25.4$.

If we check the result with respect to the rule proposed in point 7 in Section 10.2 we find that it is acceptable. The most questionable value is m_{yGE}, where the lower value is more than 40% lower than the higher value. On the other hand this moment is small compared to the other moments and this difference might be accepted. However, it will be demonstrated how the accuracy of the solution is increased by changing the coordinates for the points of intersection of the span lines.

The principles for the change of these coordinates is to move the points of intersection in such a direction that the differences are decreased. Thus e.g. point g (or f) is moved downwards in order to decrease the difference in m_{yEG}. Point g is also moved to the left in order to decrease the difference in m_{xGF}. Some of the other points are also moved in order to decrease other differences. It should be noted that points h, e and f, which were originally situated on the straight lines a-b, c-d and d-a, respectively, may be moved without taking into account these straight lines.

The following coordinates are chosen for the second analysis:

a (4.45/14.50), b (20.75/8.50), c (21.62/2.40), d (6.40/1.32), e (13.70/1.89), f (5.55/6.78), g (12.30/7.50), h (14.00/10.85).

It may be noted that the changes in coordinates are at most in the order of about 0.3 m, which is about 5% of the average span.

An analysis with the new coordinates for the points of intersection of the span lines give the following moment values:

$M_{xE} = -306$, $M_{xG} = -228$, $M_{xF} = -418$, $m_{sA1} = -47.7$, $m_{sA2} = -36.6$.

$m_{xAE} = (25.0/30.8) \Rightarrow 29.3$, $m_{xAG} = (19.1/19.8) \Rightarrow 19.6$, $m_{xEF} = (22.5/22.9) \Rightarrow 22.8$,

$m_{xGF} = (19.8/22.9) \Rightarrow 22.1$, $m_{xFC} = (30.8/36.3) \Rightarrow 34.9$.

$M_{yG} = -169$, $M_{yE} = -209$, $M_{yF} = -387$.

$m_{yDG} = (15.6/16.3) \Rightarrow 16.1$, $m_{yGE} = (7.3/7.9) \Rightarrow 7.8$, $m_{yEB} = (16.5/18.4) \Rightarrow 17.9$,

$m_{yDF} = (27.7/25.6) \Rightarrow 27.2$, $m_{yFB} = (20.3/24.9) \Rightarrow 23.8$.

A comparison with the previous analysis shows that the differences are rather small. The design might well have been based on the first analysis. The practical effect on the behaviour of the slab and the safety are certainly not noticeable.

We now know all the moment values which we need for the design of the reinforcement. Possible design moment distributions using these values are shown in Fig. 10.3.3 for reinforcement in the x-direction and support moments at the fixed edge, and in Fig. 10.3.4 for reinforcement in the y-direction. The distribution of support moments at the columns follows the general recommendations.

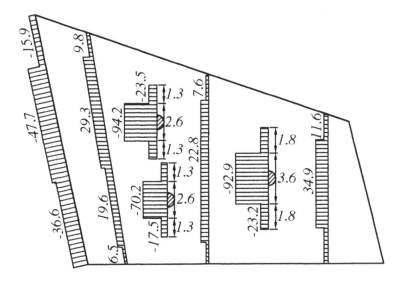

Fig. 10.3.3

In addition to the design moments shown some minimum reinforcement according to the relevant code should also be provided, as well as some corner reinforcement if prevention of top cracks is important.

181

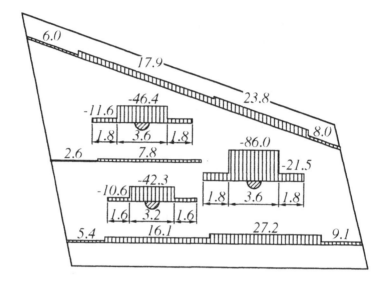

Fig. 10.3.4

We can also calculate the reaction forces on the columns. The column force is the load acting on the area within the span lines of zero shear force surrounding the column. So, for example, the area belonging to column E is 45.5 m^2 and the corresponding reaction force is 546 kN.

10.4 Edges straight and partly column supported

The treatment in this case is similar to the preceding one with the exception that reinforcement has to be arranged along the column-supported edges. Support bands are introduced along these edges, carrying the support reactions.

Where a column is situated at an obtuse corner of a slab, case b) in Fig. 10.4.1, the situation is intermediate between that of a column at a right-angled corner, case a), and a straight edge, case c). In case a) there is no negative support moment, provided that the slab is assumed to be simply supported on the column. In case c) there is a large support moment, corresponding to continuity of the slab over the column. In the intermediate case b) there may also be some support moment, particularly when the angle φ approaches 180°. The support moment increases from zero for $\varphi = 90°$ to the full support moment caused by continu-

182

ity for $\varphi = 180°$. Probably, it increases slowly at small angles and faster as φ approaches 180°. The following approximate rule is proposed for the design support moment M_s:

$$M_s = 0 \quad \text{for } \varphi < 135° \tag{10.22}$$

$$M_s = M_{s,cont}(\varphi - 135)/45 \quad \text{for } \varphi > 135° \tag{10.23}$$

where $M_{s,cont}$ is the support moment corresponding to full continuity. The reinforcement for M_s has to be well anchored.

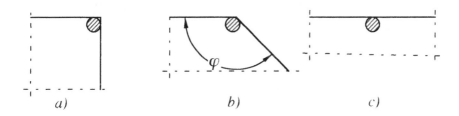

a)   *b)*   *c)*

Fig. 10.4.1

The moment M_s at an obtuse corner has different directions for the two support bands meeting at the corner. For equilibrium there must also be a moment M_b with its vector perpendicular to the bisector:

$$M_b = 2M_s\cos\varphi/2 \tag{10.24}$$

This moment may be taken by bottom reinforcement in the slab or by moment in the corner column. It is not necessary to arrange extra reinforcement to take M_b, but a check must be made that the reinforcement arranged for other reasons is sufficient to take that moment.

In order to prevent top cracks in the direction of the bisector the support reinforcement for M_s should be distributed over a certain width, selected according to the rules for support bands in Section 2.8.3. Also, where $\varphi < 135°$, some reinforcement for this purpose may be needed, preferably at right angles to the bisector.

Example 10.2

The slab in Fig. 10.4.2 is the same as in the previous example with the exception that edges B and C are supported on columns with a diameter of 0.4 m. The load is also the same. The slab is assumed to be simply supported at the columns. We can thus use the results from Example 10.1 and need only make the calculations for the support bands.

For the design calculation the circular columns are exchanged for square inscribed columns with sides 0.28 m according to Fig. 10.4.3. For the corner column an inscribed figure

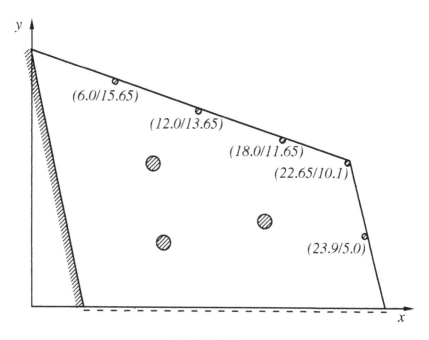

Fig. 10.4.2

is used with edges at right angles to the directions of the support bands. In this case a triangle is suitable.

The positions of the columns have been chosen so that the support band along the inside of the columns in Fig. 10.4.3 follows the supports in Example 10.1. We can thus take the load on the support bands from the lines of zero shear force in Fig. 10.3.2, assuming that the load is carried at right angles into the support band. These lines are shown in Fig. 10.4.3. In addition, there is the load on the strip between the columns, which has a width of 0.34 m, corresponding to a load of 4.0 kN/m.

The load on the support bands, including the load on the strip between the columns, is shown in Fig. 10.4.4. As the angle at L is less than 135° the bands are assumed to be simply supported at that column. The calculation of the moments can be made by means of ordinary methods according to the theory of elasticity. Here, instead, the same iterative method will be applied as for the slab according to the recommendation in Section 10.2.

We start by analysing band *H-I-J-K-L*. Points of zero shear force are assumed at the centres of all spans except span *K-L*, where this point is assumed to be at a distance of $0.625 \times 4.60 = 2.88$ m from *K*. The analysis is rather trivial and will not be shown in detail. As

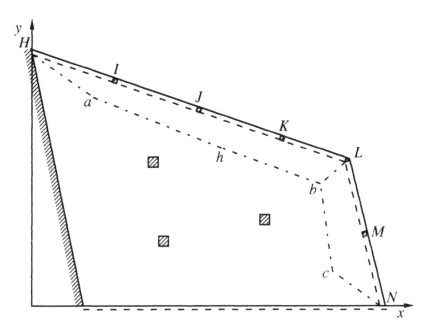

Fig. 10.4.3

an example the calculation for the part next to H will be shown. The load at the point of zero shear force in span $H-I$ is 14.2 kN/m. The equilibrium equation is

$$M_{HI} - M_H = \frac{4.0 \times 3.09^2}{2} + \frac{10.2 \times 3.09^2}{3} = 51.6 \qquad (10.25)$$

In the same way we find

$$M_{HI} - M_I = 83.8 \qquad (10.26)$$

$$M_{IJ} - M_I = 104.1 \qquad (10.27)$$

$$M_{IJ} - M_J = 109.3 \qquad (10.28)$$

$$M_{JK} - M_J = 117.5 \qquad (10.29)$$

$$M_{JK} - M_K = 120.4 \qquad (10.30)$$

185

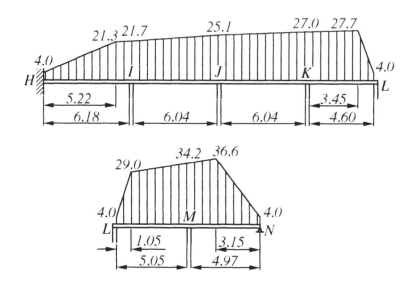

Fig. 10.4.4

$$M_{KL} - M_K = 113.6 \tag{10.31}$$

$$M_{KL} = 35.8 \tag{10.32}$$

According to the rules in Section 10.2 we find the following design moments:

$M_I = -(83.8+104.1)/3 = -62.6$, $M_J = -(109.3+117.5)/3 = -75.6$, $M_K = -(120.4+113.6)/3 = -78.0$.

$M_{HI} = 21.2$, $M_{IJ} = (41.5/33.7) \Rightarrow 39.6$, $M_{JK} = (41.9/42.4) \Rightarrow 42.3$, $M_{KL} = (35.6/35.8) \Rightarrow 35.8$. With $M_{HI} = 21.2$ we get $M_H = -30.4$ from Eq. (10.25) as the support moment at the fixed edge. This is an acceptable value although a little low. If we had used the recommendation in Section 10.2, a better value might have been $-2 \times 51.6/3 = -34.4$, giving a decrease in M_{HI} to 20.2.

All the values are acceptable according to the recommendations in Section 10.2

The maximum width of the reinforcement band is checked by means of the rules in Section 2.8.3. Thus the width on the inside of the support band may correspond to about half the average width of the element which causes the load on the band. These band widths are found to be about 0.4 m for span *H-I*, 0.8 m for span *I-J*, 0.9 m for span *J-K*, and 0.9 m for span *K-L*. To these values should be added the width of the band which is directly carried on the columns. This width is $0.20 + 0.14 = 0.34$ m. Where two different values are found on both sides of a column the average is applied.

A corresponding analysis for band $L - M - N$ with the points of zero shear force according to the recommendations at distances 3.15 and 3.11 m respectively from column M gives

$$M_{LM} = 48.5 \qquad (10.33)$$

$$M_{LM} - M_M = 156.1 \qquad (10.34)$$

$$M_{MN} - M_M = 152.2 \qquad (10.35)$$

$$M_{MN} = 28.8 \qquad (10.36)$$

$M_M = -(156.1+152.2)/3 = -102.8$, $M_{LM} = (48.5/53.3) \Rightarrow 52.1$, $M_{MN} = (49.4/28.8) \Rightarrow$ 44.3.

The difference between the two values of M_{MN} is not quite acceptable. We therefore have to move the point of zero shear force in span $M - N$ closer to M. The new distance is chosen as 2.87 m. If we only make this change we will get a greater difference in the values of M_N. We therefore also move the point of zero shear force in span $L - M$ closer to M, choosing 3.07 m.

With these new values we find

$M_M = -(148.6+135.0)/3 = -94.5$, $M_{LM} = (53.2/54.1) \Rightarrow 53.9$, $M_{MN} = (40.5/40.7) \Rightarrow 40.7$.

It will be seen that the difference between the two results is not very important. The first calculation might well have been accepted in spite of the different values of M_{MN}. The latter calculation gives a saving in reinforcement of about 5%. It may be worthwhile making a second analysis in order to save reinforcement, but it hardly influences the safety or the behaviour.

The resulting design moments in the support bands are shown in Fig. 10.4.5, given as moments in kNm/m in the band widths shown. At the obtuse corner a design moment for an extra top reinforcement is shown, which has been estimated to be about one-quarter of the moment for full continuity.

The complete design moments for the slab are the moments according to Figs 10.3.3, 10.3.4 and 10.4.5.

10.5 Edge curved and fully supported

Where the edges are curved in the analysis they may be exchanged for a circumscribed polygon, which gives a safe design. The polygon is as a rule best chosen so that each side corresponds to the supported side of an analysed element.

The general rules in Section 10.2 apply.

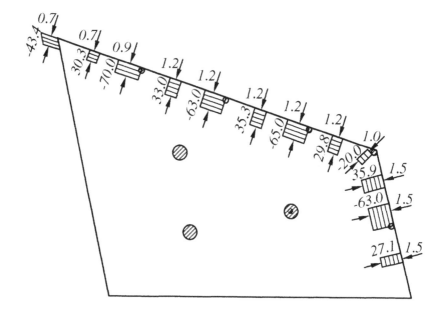

Fig. 10.4.5

Example 10.3

The slab in Fig. 10.5.1 has a simply supported edge in the shape of a quarter of an ellipse. The two straight edges are fixed. There are two square interior columns with 0.3 m sides. The load is 10 kN/m². All reinforcement is assumed to be arranged in the x- and y-directions.

The assumed circumscribed polygon and the lines of zero shear force are shown in Fig. 10.5.2. It may be noted that each side of the polygon forms the side of an element where the other sides are lines of zero shear force. The pattern of lines of zero shear force has been determined according to the rules in Section 10.2 and adjusted after a first analysis, which is not shown here, in order to reduce the differences between values of span moments.

The coordinates of the points of intersection between the span lines of zero shear force are:

a (2.8/8.6), b (10.0/7.3), c (15.7/5.3), d (17.6/1.5), e (9.0/2.4), f (2.8/3.2).

Performing the analysis in the same way as demonstrated in Example 10.1 we find the following relations:

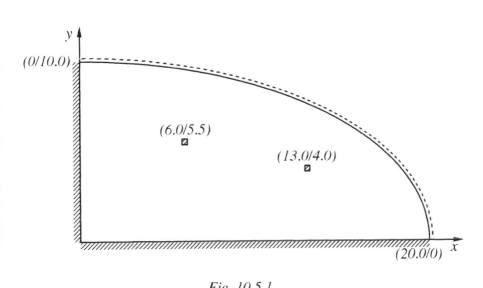

Fig. 10.5.1

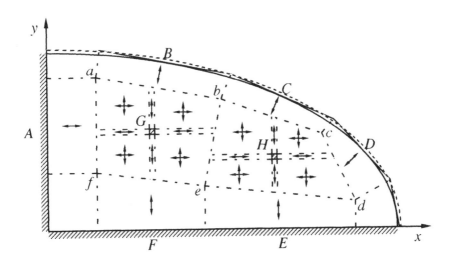

Fig. 10.5.2

$$m_{xAG} - m_{xA} = 39.2 \tag{10.37}$$

$$5.24 m_{xAG} - M_{xG} = 248.7 \tag{10.38}$$

$$5.23 m_{xGH} - M_{xG} = 289.3 \tag{10.39}$$

$$4.30 m_{xGH} - M_{xH} = 273.3 \tag{10.40}$$

$$4.22 m_{xHD} - M_{xH} = 232.6 \tag{10.41}$$

$$m_{HD} = 14.0 \tag{10.42}$$

$$m_{yFG} - m_{yF} = 39.5 \tag{10.43}$$

$$6.80 m_{yFG} - M_{yG} = 211.9 \tag{10.44}$$

$$6.86 m_{yGB} - M_{yG} = 194.0 \tag{10.45}$$

$$m_{GB} = 11.6 \tag{10.46}$$

$$m_{yEH} - m_{yE} = 19.4 \tag{10.47}$$

$$7.13 m_{yEH} - M_{yH} = 146.3 \tag{10.48}$$

$$6.92 m_{yHC} - M_{yH} = 153.1 \tag{10.49}$$

$$m_{HC} = 8.2 \tag{10.50}$$

Applying the rules in section 10.2 and with the notation used in Example 10.1 we further get

$m_{xA} = -26.1$, $m_{yF} = -26.3$, $m_{yE} = -12.9$.

$M_{xG} = -179.3$, $M_{xH} = -168.6$, $M_{yG} = -135.3$, $M_{yH} = -99.8$.

$m_{xAG} = (13.1/13.2) \Rightarrow 13.2$, $m_{xGH} = (21.0/24.3) \Rightarrow 23.5$, $m_{xHD} = (15.2/14.0) \Rightarrow 14.9$,

$m_{yFG} = (13.2/11.3) \Rightarrow 12.7$, $m_{yGB} = (8.6/11.6) \Rightarrow 10.9$, $m_{yEH} = (6.5/6.5) \Rightarrow 6.5$,

$m_{yHC} = (7.7/8.2) \Rightarrow 8.1$.

It should be noted that the values for m_{xHD}, m_{yGB} and m_{yHC} are only valid if $m_x = m_y$ within the areas close to the supports, cf. Eq. (10.1). It is economically more efficient to have different values of m_x and m_y. Thus to reinforce for m_{GB} we may use a value of m_x which only corresponds to the reinforcement designed for the x-direction. With the rules used for moment distribution, this value may be taken as the lower of one-third of m_{xAG} or m_{xGH}, which is $13.2/3 = 4.4$. At B, $\sin^2\varphi = 0.035$. Applying Eq. (10.1) we find

$$0.035 \times 4.4 + 0.965 m_{yGB} = 11.6 \tag{10.51}$$

giving $m_{yGB} = 11.9$ for that element and $m_{yGB} = (8.6/11.9) \Rightarrow 11.1$.
Applying the same principles for m_{HD} and m_{HC} we find

$$0.545 m_{xHD} + 0.455 \times 2.2 = 14.0 \qquad (10.52)$$

giving $m_{xHD} = (15.2/23.9) \Rightarrow 21.7$, and

$$0.167 \times 7.8 + 0.833 m_{yHC} = 8.2 \qquad (10.53)$$

giving $m_{yHC} = (7.7/8.3) \Rightarrow 8.2$.

It is only m_{xHD} which is appreciably influenced by this analysis. The reason is that the angle φ between the edge and the x-axis is close to 45° in this case. Where this angle is closer to 0° or 90° the influence is smaller.

Figs 10.5.3 and 10.5.4 show possible distributions of design moments for reinforcement in the x- and y-directions respectively. The general recommendations regarding distribution have been followed. The width over which the support reinforcement in the x-direction has been distributed is approximately equal to the length of the relevant support line of zero shear force in Fig. 10.5.2. The distribution width for the support reinforcement in the y-direction has been chosen the same as in the x--direction, cf. Section 8.1.5.

The widths of distribution of span reinforcement are based on the pattern of span lines of zero shear force in Fig. 10.5.2 and are always chosen to be on the safe side.

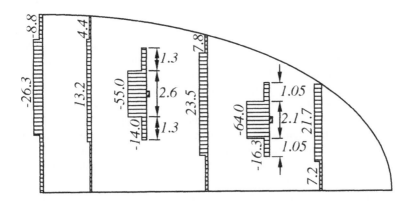

Fig. 10.5.3

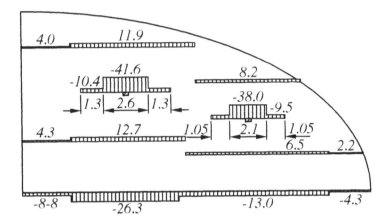

Fig. 10.5.4

10.6 Edge curved and column supported

Example 10.4

The slab in Fig. 10.6.1 is the same as in the previous example with the exception that the curved edge is supported by square columns 0.3m×0.3 m situated with their inner edges at the line of support in that example. The coordinates of the centres of the columns are given. The slab is assumed to be simply supported on the columns.

The pattern of lines of zero shear force inside the slab is taken unchanged from the previous example. Fig. 10.6.2 shows the span lines and also the support bands which carry the support reactions to the columns, shown as dashed lines between column corners. The support bands are assumed to meet the fixed edges 0.3 m from the corners.

The shapes of the elements closest to the free edge are not exactly the same as in the previous example. So, for example, the moment on span line *a-b* comes from two elements with slightly different directions of the support lines. The moment to the left of *B* is $m = 8.9$. With $\sin^2\varphi = 0.006$ for that support line we also find $m_y = 8.9$. For the part to the right of *B* we find $m = 9.0$. With $\sin^2\varphi = 0.062$ and $m_x = 4.4$ (cf. Eq. (10.52)) we get $m_y = 9.0$. The average on line *a-b* is $m_{yGB} = 9.0$ and using the other value from the previous example we get $m_{yGB} =$

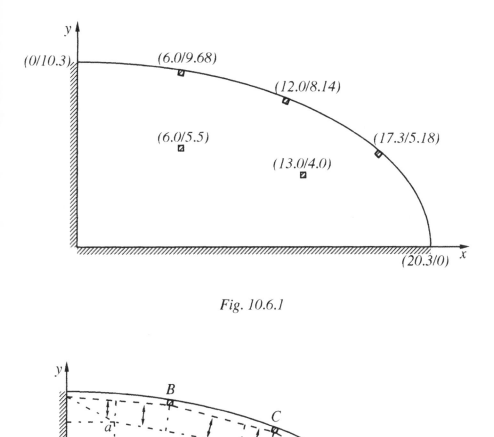

Fig. 10.6.1

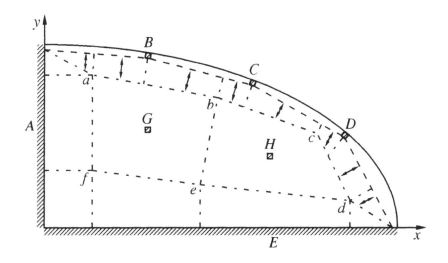

Fig. 10.6.2

$(8.6/9.0) \Rightarrow 8.9$. In the same way we find $m_{yHC} = (7.7/5.4) \Rightarrow 7.1$ and $m_{xHD} = (15.2/8.8) \Rightarrow 13.6$. These values are slightly lower than in the previous example, which may be expected as the dimensions of the elements are smaller.

The load on the support bands corresponds to the load on the elements outside the line *a-b-c-d* supplemented by the lines from *a* and *d* to the intersections between the support bands and the supports. The load is assumed to be carried at right angles to the support bands. The resulting load distribution on the support band is shown in Fig. 10.6.3, where a series of linear approximations is used.

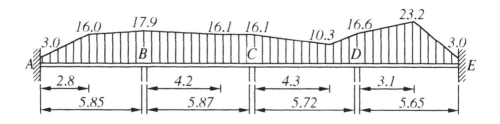

Fig. 10.6.3

The calculation of moments in the support bands can be performed according to the theory of elasticity or in the same way that is used for the elements inside the slab, i.e. with assumed positions of points of zero shear force. In both cases the support moments should be reduced according to the rules in Eqs (10.22) and (10.23). The method with assumed lines of zero shear force will be demonstrated.

The angles between meeting support bands are 170.0° at *B*, 165.3° at *C* and 147.5° at *D*. The corresponding reduction factors for support moments are 0.778 at *B*, 0.673 at *C* and 0.278 at *D*.

We can start by assuming that all points of zero shear force are at the centres of the spans, even though it can easily be estimated that these are not the correct positions. Simple statics then gives the following relations

$$M_{AB} - M_A = 48.3 \tag{10.54}$$

$$M_{AB} - M_B = 71.2 \tag{10.55}$$

$$M_{BC} - M_B = 73.6 \tag{10.56}$$

$$M_{BC} - M_C = 69.9 \tag{10.57}$$

$$M_{CD} - M_C = 55.2 \tag{10.58}$$

$$M_{CD} - M_D = 47.5 \tag{10.59}$$

$$M_{DE} - M_D = 82.5 \tag{10.60}$$

$$M_{DE} - M_E = 70.2 \tag{10.61}$$

Applying the recommendations in Section 10.2 and the reduction factors given above we can calculate the following moments:

$M_A = -48.3 \times 2/3 = -32.2$, $M_B = -0.778 \times (71.2 + 73.6)/3 = -37.6$,
$M_C = -0.673 \times (69.9 + 55.2)/3 = -28.1$, $M_D = -0.278 \times (47.5 + 82.5)/3 = -12.0$,
$M_E = -70.2 \times 2/3 = -46.8$.
$M_{AB} = (16.1/33.6) \Rightarrow 29.2$, $M_{BC} = (36.0/41.8) \Rightarrow 40.4$, $M_{CD} = (27.1/35.5) \Rightarrow 33.4$,
$M_{DE} = (70.5/23.4) \Rightarrow 58.7$.

There are unacceptably large differences between the two values of span moments for M_{AB} and for M_{DE}. The positions of the points of zero shear force are changed in these two spans, while the positions are kept unchanged in the other two spans for the next calculation. An estimate by means of Eq. (10.2) shows that it may be suitable to move the point of zero shear force in span AB 0.26 m to the right and in span DE 0.35 m to the left. After this change we get:

$M_A = -43.1$, $M_B = -34.5$, $M_C = -28.1$, $M_D = -10.1$, $M_E = -62.3$.
$M_{AB} = (21.5/24.8) \Rightarrow 24.0$, $M_{BC} = (39.1/41.8) \Rightarrow 41.1$, $M_{CD} = (27.1/37.4) \Rightarrow 34.8$, $M_{DE} = (51.8/31.2) \Rightarrow 46.7$.

The difference in span moments is still too large in span DE. The point of zero shear force is moved 0.28 m further to the left. At the same time the point of zero shear force in span CD is moved 0.12 m to the right in order to decrease the difference in that span. After these changes we get the moments:

$M_A = -43.1$, $M_B = -34.5$, $M_C = -29.1$, $M_D = -8.5$, $M_E = -75.7$.
$M_{AB} = (21.5/24.8) \Rightarrow 24.0$, $M_{BC} = (39.1/40.8) \Rightarrow 40.4$, $M_{CD} = (30.6/35.0) \Rightarrow 33.9$,
$M_{DE} = (39.3/37.8) \Rightarrow 38.9$.

These values are quite acceptable and may be used for the design.

The difference in direction of moment vectors on both sides of the columns gives rise to a moment according to Eq. (10.24), which has to be balanced by reinforcement. Thus at B, for instance, this moment is $2 \times 34.5 \cos(170/2) = 6.0$ kNm. At C the moment is 7.5 and at D 4.8. These moments should be taken by bottom reinforcement at right angles to the edge.

Fig. 10.6.4 shows the moments in the support band. The values are for M in kNm, not for m in kNm/m as in most other corresponding figures. The moments have been shown distributed on a width of about 0.6 m, which may be suitable. Alternative widths and positions are of course also possible. At the columns it has been indicated that there are two different directions of the reinforcement corresponding to the support moments. Two different systems of reinforcement bars may be arranged, which should be well anchored, or bars may be horizontally bent over the columns. In the latter case some secondary reinforcement might be needed to avoid the risk of splitting due to the radial pressure against the bend.

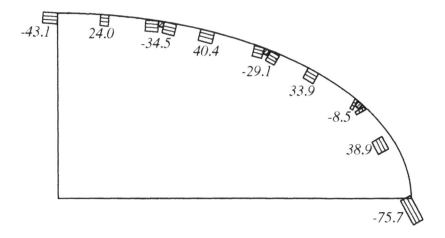

Fig. 10.6.4

10.7 Slab cantilevering outside columns

Where the slab cantilevers outside columns at least two different approaches may be used. Either use can be made of support bands which support cantilevers approximately at right angles to the support bands, or the treatment can be based on the design procedure with reinforcement in two directions, described in Section 10.2. In Example 10.5 both approaches are discussed and an analysis with the latter approach is demonstrated, whereas in Example 10.6 three different approaches are demonstrated in detail and compared

It must be noted that the approach according to Section 10.2 does not automatically give results on the safe side, as the rules for acceptable moment distributions may be clearly violated. In particular, there may be some negative cantilever moments which are underestimated with this method, leading to a risk of uncontrolled cracking. This will be commented upon in Example 10.6. However, the safety against collapse will probably always be adequate if the recommendations are followed.

A particular problem which has to be taken into account is the case of reinforcing bars, normally bottom bars, which reach a free edge soon after crossing a line of zero shear force. As the reinforcement is assumed to be able to yield at such a line there needs to be a certain distance to the free edge, where the steel stress is zero. In order to make sure that the reinforcement is fully efficient in the lines of zero shear force it is recommended that the following rule should be applied:

A reinforcing bar is only taken into account in the design if it has a length from the line of zero shear force to the point where it is terminated at a free edge which is at least equal to l/8, where l is the length of the span where the bar is active.

This rule is not very precise, as it is not always possible to define the relevant value of *l*, but it can be expected to give results on the safe side in most situations. In case of doubt a value on the safe side is chosen.

Example 10.5

The slab in Fig. 10.7.1 has columns placed as in the previous example, but the slab cantilevers 1.5 m outside the columns which stood along the edge in that example. The columns have the same centres, but have their sides along the directions of the coordinate axes. All columns have a square section with sides 0.3 m. The load is 10 kN/m^2, as in the previous examples. The free edge has the shape of an ellipse.

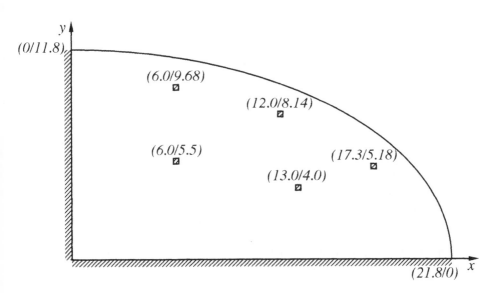

Fig. 10.7.1

Fig. 10.7.2 shows the principles of the approach with a support band along the outer row of columns. The treatment is very similar to that in Example 10.4 and the numerical calculations will not be shown. The main difference from Example 10.4 is that the cantilevering parts give rise to negative moments over the support band (corresponding to top reinforcement perpendicular to the free edge), which have to be taken into account in writing the equilibrium equations for the one-way elements inside the support band. It is recommended that the top reinforcement is arranged with a certain concentration towards the columns. The support band will take a substantially higher load than in Example 10.4.

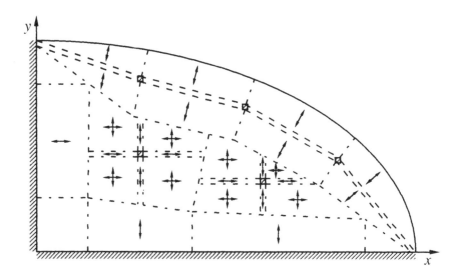

Fig. 10.7.2

Fig. 10.7.3 shows lines of zero shear force when the slab is designed according to Section 10.2, with all reinforcement parallel to the coordinate axes. The system of lines of zero shear force has been determined according to the principles in Section 10.2. After some adjustments of the positions the following coordinates of the points of intersection have been chosen:

a (2.80/7.80), b (8.80/7.80), c (9.60/5.40), d (14.60/6.60), e (16.10/1.80), f (9.00/2.40), g (2.80/3.20), h (3.00/11.69), i (9.45/10.63), j (15.51/8.29), k (20.76/3.60).

When establishing the equilibrium relations for the elements next to the free edge it is not evident how much of the span (bottom) reinforcement should be taken into account. To be on the safe side, it has been assumed that all the load is carried by the support reinforcement in cases where there is a small angle between the support line and the edge, which is

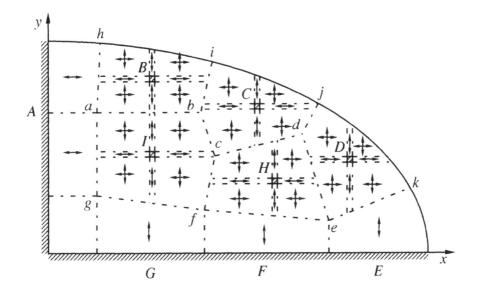

Fig. 10.7.3

the case for reinforcement in the x-direction at column D and in the y-direction for all three columns B, C and D. This is done in order to eliminate the risk of too little top reinforcement in the cantilevering part.

For the elements where a free edge forms a small angle with the reinforcement direction the recommendation above has been applied. The relevant cases are reinforcement in the x-direction near points h, i and j. In all three cases the value of l can be taken to be approximately 6.0 m and the distance from the line of zero shear force to the free edge not more than about 0.75 m. Depending on the slope of the free edge the reduction in width on which the reinforcement is active is found to be 0.06 m at h, 0.20 m at i and 0.36 m at j. These reductions have been used for the determination of the active widths for the m-values in the equations below.

The following equilibrium relations have been found:

$$3.83m_{xAB} - 4.00m_{xAI} = 165.3 \tag{10.62}$$

$$m_{xAI} - m_{xA2} = 39.2 \tag{10.63}$$

$$3.83m_{xAB} - M_{xB} = 164.3 \tag{10.64}$$

$$4.99m_{xAI} - M_{xI} = 220.0 \tag{10.65}$$

199

$$2.63 m_{xBC} - M_{xB} = 138.3 \tag{10.66}$$

$$5.03 m_{xIH} - M_{xI} = 256.6 \tag{10.67}$$

$$5.03 m_{xBC} - M_{xC} = 178.9 \tag{10.68}$$

$$4.11 m_{xIC,IH} - M_{xH} = 211.4 \tag{10.69}$$

$$1.92 m_{xCD} - M_{xC} = 106.2 \tag{10.70}$$

$$4.20 m_{xHD} - M_{xH} = 115.8 \tag{10.71}$$

$$5.72 m_{xCD,HD} - M_{xD} = 114.6 \tag{10.72}$$

$$-M_{xD} = 97.2 \tag{10.73}$$

$$m_{yGI} - m_{yG} = 39.5 \tag{10.74}$$

$$6.80 m_{yGI} - M_{yI} = 211.9 \tag{10.75}$$

$$6.72 m_{yIB} - M_{yI} = 144.2 \tag{10.76}$$

$$6.31 m_{yIB} - M_{yB} = 91.3 \tag{10.77}$$

$$-M_{yB} = 67.0 \tag{10.78}$$

$$m_{yFH} - m_{yF} = 22.2 \tag{10.79}$$

$$6.17 m_{yFH} - M_{yH} = 104.3 \tag{10.80}$$

$$6.02 m_{yHC} - M_{yH} = 96.9 \tag{10.81}$$

$$6.51 m_{yHC} - M_{yC} = 114,1 \tag{10.82}$$

$$-M_{yC} = 68.5 \tag{10.83}$$

$$4.66 m_{yED} - 5.70 m_{yE} = 207.0 \tag{10.84}$$

$$4.63 m_{yED} - M_{yD} = 146.8 \tag{10.85}$$

$$-M_{yD} = 85.2 \tag{10.86}$$

From these relations we can calculate the moments according to the recommendations in Section 10.2:

$m_{xAI} = -27.8$ (for $-m_{xAI} = 2m_{xAB}$), $m_{xA2} = -26.1$, $M_{xB} = -(164.3 + 138.3)/3 = -100.9$,

$M_{xI} = -158.9, M_{xC} = -95.0, M_{xH} = -109.1, M_{xD} = -97.2, m_{yG} = -26.3, M_{yI} = -118.7,$
$M_{yB} = -67.0, m_{yF} = -14.8, M_{yH} = -67.1, M_{yC} = -68.5, m_{yE} = -25.8, M_{yD} = -85.2.$
$m_{xAB} = (14.1/16.6) \Rightarrow 16.0, m_{xAI} = (13.1/13.2) \Rightarrow 13.2, m_{xBC} = (14.2/16.7) \Rightarrow 16.1,$
$m_{xIC} = (19.4/21.2) \Rightarrow 20.8, m_{xIH} = (19.4/24.9) \Rightarrow 23.5, m_{xCD} = (5.8/3.0) \Rightarrow 5.1,$
$m_{xHD} = (1.6/3.0) \Rightarrow 2.7, m_{yGI} = (13.2/13.7) \Rightarrow 13.6, m_{yIB} = (3.8/3.9) \Rightarrow 3.9,$
$m_{yFH} = (7.4/6.0) \Rightarrow 7.1, m_{yHC} = (5.0/7.0) \Rightarrow 6.5, m_{yED} = (12.9/13.3) \Rightarrow 13.2.$

Fig. 10.7.4 shows possible distributions of design moments for reinforcement in the x-direction and Fig. 10.7.5 for the y-direction. For the reinforcement in the x-direction some simplifications have been used for m_{xA}, where the larger value m_{xAI} only has been used, for m_{xAB} and m_{xAI}, where the value 16.0 has been used for both, and for m_{xBC}, m_{xIC} and m_{xIH}, where the value 22.0 has been used for all three. The chosen values are somewhat arbitrary, but on the safe side. Some corresponding simplifications might have been made also for the y-direction, but the differences there are greater.

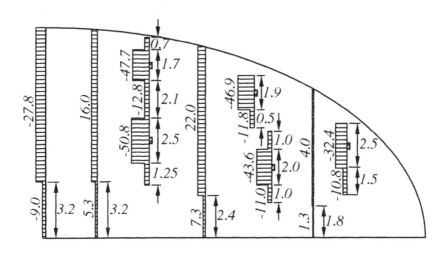

Fig. 10.7.4

The support moment at E has been given a concentration towards the end where it meets the free edge. The reason is that it has been estimated that the slab has a small cantilever action in that part.

The support moments over the columns have been distributed according to the general recommendations on approximately the width of the corresponding support lines of zero shear force. It is, however, only meaningful to arrange reinforcement where there is a suffi-

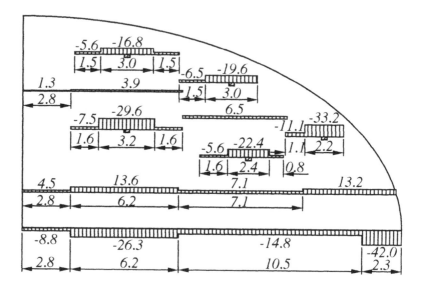

Fig. 10.7.5

cient length of reinforcing bar. This is the reason why the distribution is unsymmetrical for M_{xD}, M_{yC} and M_{yD}.

In addition to the reinforcement for the design moments at least one top bar and one bottom bar should be placed along the free edge. At least some of the top reinforcement should be bent around these bars according to the principle illustrated in Fig. 2.10.2.

Example 10.6

The triangular slab in Fig. 10.7.6 has all edges free, and is supported on three square columns $0.3\,\text{m} \times 0.3\,\text{m}$. It is symmetrical with respect to the x-axis. The positions of the corners and the centres of the columns are given as coordinates. The load is $9\,\text{kN/m}^2$.

Three different possibilities for designing this slab will be discussed. In all the solutions use will be made of the symmetry so that only the upper half of the slab is analysed.

Approach 1

The first approach is a direct application of the design procedure according to Section 10.2, with all reinforcement parallel to the coordinate axes. The corresponding pattern of lines of zero shear force is shown in Fig. 10.7.7. The only points which have to be chosen are *a* and

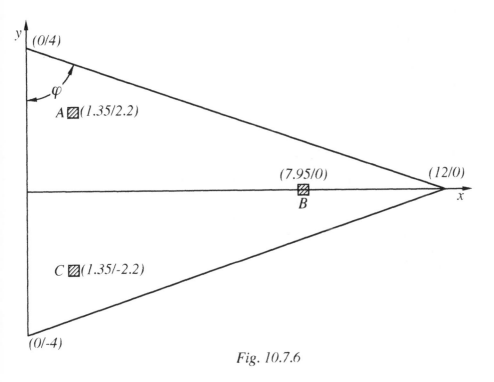

Fig. 10.7.6

b. According to the recommendations the line between these points has a direction approximately at right angles to the line between columns *A* and *B*. The positions have been determined so that the span moments m_{xf} from the two elements on both sides of the line are approximately equal. In this way suitable coordinates have been found to be:

a (4.8/2.4), *b* (4.0/0).

The support line of the element to the left of column *A* has a length of 3.6 m. Applying Eq. (2.5) we find

$$M_{xsA} = -\frac{9 \times 1.2^2 (3.6 + 2 \times 4.0)}{6} = -25.06 \qquad (10.87)$$

The support line of the element to the right of column *A* has a length of 3.5 m. As the span reinforcement meets the edge at a skew angle the recommendation on page 197 is applied. The relevant length *l* is 6.3 m in this case and the minimum length of reinforcing bars from the line of zero shear force to the edge is about 0.8 m. With the slope $dy/dx = -1/3$ the reduction in active width is 0.27 m and the active width 2.4-0.27 = 2.13 m. From Eq. (2.6) we then find

203

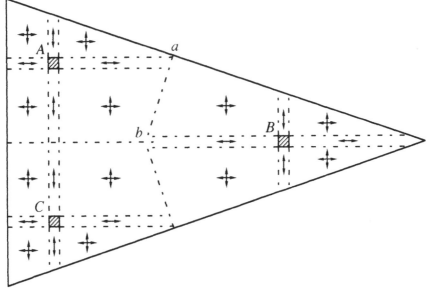

Fig. 10.7.7

$$2.13\, m_{xf} - M_{xsA} = \frac{9}{6}[3.3^2 \times 1.1 + (3.3^2 + 3.3 \times 2.5 + 2.5^2)2.4] = 109.37 \quad (10.88)$$

As we analyse the upper half of the slab we include half the widths of the one-way elements in the x-direction at column B. For the element to the right of column B we then find that the length of the relevant support line is 1.3 m. From Eq. (2.4) we find

$$M_{xsB} = -\frac{9 \times 3.9^2}{6} \times 1.3 = -29.66 \quad (10.89)$$

It should be noted that this is only half the moment above the column.

For the element to the left of column B the length of the support line is 1.4 m. The active width of the span reinforcement is taken to have the same value as above, 2.13 m. We find from Eq. (2.6)

$$2.13\, m_{xf} - M_{xsB} = \frac{9}{6}[(3.8^2 + 3.8 \times 3.0 + 3.0^2)2.4 - 3.0 \times 1.0] = 111.92 \quad (10.90)$$

From Eqs (10.87) - (10.90) we find $m_{xf} = (39.58/38.62) \Rightarrow 39.34$.

204

The length of the support line of the element above column A is 4.78 m. From Eq. (2.6) we get

$$M_{ysA} = -\frac{9}{6}[\,(\,1.65^2 + 1.65 \times 0.05 + 0.05^2)\,4.8 - 0.05^2 \times 0.02\,] = -20.21 \qquad (10.91)$$

The length of the support line of the element below column A is 4.68 m. From Eq. (2.5):

$$4.68\,m_{yf} - M_{ysA} = \frac{9 \times 2.05^2\,(4.68 + 2 \times 4.0)}{6} = 79.93 \qquad (10.92)$$

and so $m_{yf} = 12.76$.

The length of the support line of the element above column B is 7.50 m. From Eq. (2.4):

$$M_{ysB} = -\frac{9 \times 2.25^2}{6} \times 7.50 = -56.95 \qquad (10.93)$$

We now know all the design moments and just have to choose a suitable distribution according to the general recommendations. Such a distribution can be found in Fig. 10.7.10 below.

Approach 2

In this approach use is made of a support band between columns A and B, cantilevering to the left of A, Fig. 10.7.8. The support band is divided into two bands, each passing through a corner of each column and passing through the column section. In this way the static system becomes clear. A somewhat more economical assumption might have been to let the bands rest on column corners without passing through the column sections.

The part to the right of column B is assumed to be balanced by a corresponding part to the left of the column. The loads within these areas are taken by reinforcement in the x-direction. In the remaining part of the slab the load is assumed to be primarily carried by reinforcement in the y-direction in strips which rest on the support bands.

It would have been somewhat more economical and statically correct to use reinforcement perpendicular to the edges to take the moments in the cantilevering parts of the slab. This is possible, but it leads to a more complicated static system because the strips which have to balance each other inside the slab will form an angle with each other, causing a resulting moment which has to be taken into account.

The moment M_{xsB} can be taken from the calculation in Approach 1 above, and is thus −29.66 on the upper half of the slab.

The support reaction from the triangular elements on the one-way strip in the y-direction is $9 \times 6 \times (1.3 - y)$ for $y > 0.15$ m. This gives a moment of −13.69 kNm. To this should be added the influence of the direct load on the one-way strip, which gives a moment of approximately $-9 \times 0.3 \times 1.2^2/2 = -3.52$. The total moment is thus $M_{ysB} = -17.21$.

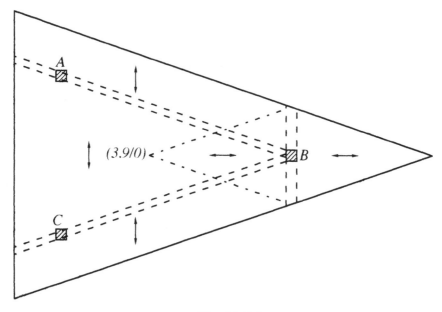

Fig. 10.7.8

The load on the whole slab is primarily carried in the y-direction for x<3.9. For 3.9<x<7.8 the load is primarily carried in the y-direction for y>(x–3.9)/3. The cantilever moments in these strips are

$$m_{ys} = -9 \times 1.25^2/2 = -7.03 \quad \text{for } 0<x<6.075 \tag{10.94}$$

$$m_{ys} = -(15.9 - 2x)(x - 4.2) \quad \text{for } 6.075<x<7.8 \tag{10.95}$$

The positive moments in the y-strips are

$$m_{yf} = 9(2.55 - x/3)^2/2 - 7.03 \quad \text{for } 0<x<3.9 \tag{10.96}$$

$$m_{yf} < 0 \quad \text{for } x>3.9 \tag{10.97}$$

The maximum value of m_{yf} is 22.23 kNm/m and the total moment is 28.90 kNm. The loads on the support bands are

$$Q\sin\varphi = 9(4 - x/3) = 36 - 3x \quad \text{for } 0<x<3.9 \tag{10.98}$$

$$Q\sin\varphi = 9(5.3 - 2x/3) = 47.7 - 6x \quad \text{for } 3.9 < x < 7.8 \tag{10.99}$$

The analysis of the bands is simple in principle, as they are statically determinate with zero moment at B, but the analysis is rather lengthy. The result is a support moment $M_A = -26.41$ and a maximum span moment $M_f = 87.31$ at $x = 4.4$.

A distribution of design moments based on this approach can be found in Fig. 10.7.10.

Approach 3

This approach follows the recommendations in Section 10.2 with the exception that the two reinforcement directions are not at right angles but are instead parallel to the free edges in the main part of the slab. One direction is parallel to the y-axis. The other direction is different in the upper and lower half of the slab, as it follows the nearest free edge. The upper half will be analysed and the reinforcement directions there follow the z- and y-axes in Fig. 10.7.9, which also shows the pattern of lines of zero shear force. Above column B, however, the reinforcement directions follow the directions of the x- and y-axes in order to avoid having three crossing layers of reinforcement. As the elements to the right of that column carry all the load as a cantilever the load-bearing directions in these elements follow the x- and y-directions, as indicated in the figure.

The theoretical support area is taken as the inscribed area with the sides parallel to the reinforcement directions. Other approaches are also possible, which may decrease the moments somewhat, but the difference is small and the chosen approach is in accordance with the general recommendations.

It is possible to use directions of support reinforcement over columns A and C which are at right angles to the edge in one or both directions. This would mean a small reduction in total reinforcement, but would have the disadvantages that the reinforcement would not be as active in controlling cracks at right angles to the edge and would not reach the very corner.

In applying the formulas from Section 2.3 there are two possibilities when the reinforcement directions are not at right angles. One possibility is to use the c-values in the direction of the relevant reinforcement and the l-values at right angles to that direction. Another possibility is to take the moment equation around the support line and use the relevant components of the design moments to balance this moment. Both approaches require that the other reinforcement direction is parallel to the support line. This requirement is not fulfilled at column B, where three different reinforcement directions meet. This problem will be discussed later.

It can be established that suitable coordinates for points c and d are the same as for a and b in Approach 1, thus

c (4.8/2.4), d (4.0/0).

For the element to the left of column A it is simplest to take the moment around the support line. The value of the moment is exactly the same as in Approach 1 above, so we can use that value.

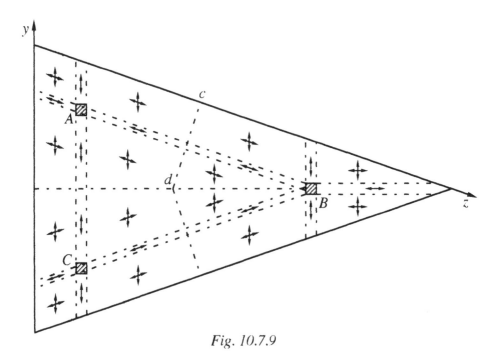

Fig. 10.7.9

$$M_{zsA}\sin\varphi = -25.06 \qquad (10.100)$$

With $\sin\varphi = 0.9487$ we get $M_{zsA} = -26.41$, which is the same value as the support moment in the support band in Approach 2.

For the element to the right of column A we can use the same approach. The active width of the span reinforcement is taken as 2.5 m.

$$(2.5m_{zf} - M_{zsA})\sin\varphi = 109.37 \qquad (10.101)$$

which gives $m_{zf} = 35.55$.

The support moment is $M_{xsB} = -29.66$ on the width of the upper half of the slab just as in the previous approaches. For the element to the left of column B we may again take the moment around the vertical support line and make use of the value from Approach 1. It should be noted that m_{zf} and M_{xsB} have different directions.

$$2.5m_{zf}\sin\varphi - M_{xsB} = 111.92 \qquad (10.102)$$

which gives $m_{zf} = 34.68$. Using also the value 35.55 from above we can, according to the recommendation, calculate the design value $m_{zf} = 35.33$.

In calculating the moments for the reinforcement in the y-direction for the element above column A it is simplest to take the c-value in the y-direction and the l-values in the x-direction. The projected length in the x-direction of the support line is 4.43 m. From Eq. (2.5) we get

$$M_{ysA} = -\frac{9 \times 1.25^2 (4.43 + 2 \times 4.8)}{6} = -32.88 \qquad (10.103)$$

For the element below column A in the figure, the projected length of the support line is 4.37 m. Assuming that the effective width of the span reinforcement is 4.0 m, Eq. (2.6) gives

$$4.0 m_{yf} - M_{ysA} = \frac{9}{6} [(2.55^2 + 2.55 \times 1.22 + 1.22^2) 4.0 + 1.22^2 \times 0.37] \qquad (10.104)$$
$$= 67.44$$

from which $m_{yf} = 8.64$.

At column B the situation is somewhat complicated for calculating the moment for the reinforcement in the y-direction, as there are moments corresponding both to the x- and z-directions. We can start by analysing the influence of the load on the triangle to the right of the line through the left side of the column (this triangle is composed of the one-way strip and the corner-supported triangle), which gives no problem. This gives rise to the moment

$$M_{1ysB} = -\frac{9 \times 1.25^2}{6} \times 3.75 = -8.79 \qquad (10.105)$$

For the trapezoidal element to the upper left of column B we take the moment around the support line. This equilibrium is influenced by the part of the moment M_{xsB} which acts along the side of the element. It is not quite clear how this moment influences the equilibrium, as some of the corresponding reinforcement also passes the support line around which the equilibrium is studied, and helps in taking some of the support moment in this line. A comparison can be made with the solution according to Approach 2, where the element to the right of column B is balanced by a corresponding element to the left. In that case evidently some reinforcement in the x-direction passes the support band and helps in taking a support moment around the support band.

An assumption on the safe side is to assume that the reinforcement in the x-direction does not take any part in carrying the support moment around the support line. The part of M_{xsB} which has to be taken into account comes from the triangle to the upper right of the column and is $\Delta M_{xsB} = -1.15 \times 9 \times 3.45^2/6 = -20.53$. In applying Eq. (2.5) around the support line we note that $c = 1.25\sin\varphi$, $l = 3.37/\sin\varphi$ and $l_1 = 3.0/\sin\varphi$:

$$M_{2ysB}\sin\varphi = -\frac{9 \times 1.25^2 3.37 + 2 \times 3.0 \sin\varphi}{6} - 20.53\cos\varphi \qquad (10.106)$$

This gives $M_{2ysB} = -28.80$. Adding the two parts we get $M_{ysB} = -8.79 - 28.80 = -37.59$. There is no positive span moment m_{yf} in the element to the left of column B.

A possible distribution of design moments is shown in Fig. 10.7.10.

Fig. 10.7.10 shows the proposed distributions of design moments according to the three different approaches. As well as the reinforcement for the design moment there should also be some reinforcement to minimize the risk of large cracks. Thus, there should be some reinforcement along the free edges where such reinforcement is not given by the design moments. Only Approach 3 gives such reinforcement along the sloping edge in the figure, but this approach does not give any such reinforcement beside column B. It may also be appropriate to put minimum reinforcement in some areas where the design moments do not require reinforcement.

Comparison

A comparison between the results shows that Approach 1 gives rather little support reinforcement in the y-direction in the left-hand part of the slab and much in the right-hand part and leaves a part in between without support reinforcement. It also gives span reinforcement in the x-direction which is too unsymmetrical with respect to column A. For both these reasons, some additional reinforcement has to be used in order to minimize the risks of cracking and maybe even to ensure adequate safety. The solution given by Approach 1 is not to be recommended, at least not without introducing some additional reinforcement, based on an estimate of the behaviour of the slab. One reason why the approach is not to be recommended has to do with the fact that there is not much redundancy in the structure, as the support moments are statically determinate, unlike in the previous examples.

When comparing the results of Approaches 2 and 3, which both certainly give adequate safety, it may be noted that Approach 3 gives a reinforcement distribution with a certain concentration around the columns, but also with some reinforcement over a greater width. Approach 2 gives a heavy concentration along the support band but no concentration in the y-direction at column A. If this type of solution is used in design it is recommended not to follow this theoretical distribution strictly, but to redistribute some of the reinforcement in the support band over a larger width and to concentrate some of the support reinforcement in the other direction over the columns.

It would seem that Approach 3 is, in the first place, to be recommended.

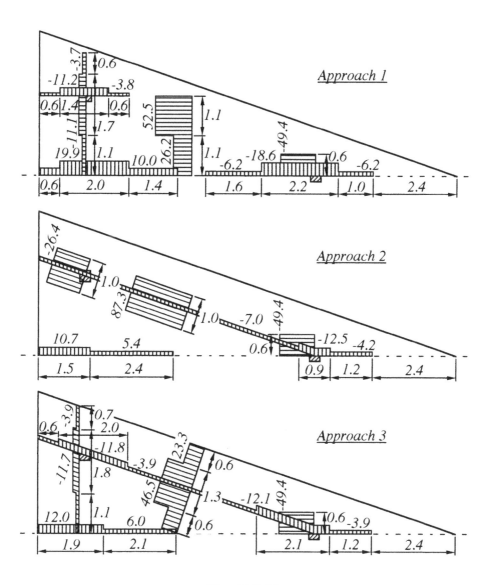

Fig. 10.7.10

211

CHAPTER 11 — *L-shaped slabs and large wall openings*

11.1 General

The slabs in this chapter can be treated by means of support bands, but the approach used here with corner-supported elements is simpler, more systematic and safer, as the well-established rules for such elements can be applied.

However, the static system is more complicated than in a regular flat slab, for example. There is no element to take the support moment from the corner-supported element, but this support moment has to be transferred to the wall in the direction of the reinforcement. Thus for element *3* in Fig. 11.1.1 the forces in the support reinforcement in the x-direction have to be balanced by a rather complicated moment distribution within element *2*, carrying the forces to the support for element *2*. It has been demonstrated in *Strip Method of Design* that it is possible to find such complete moment distributions, which are on the safe side, but that it is sufficient to analyse only the elements shown in Fig. 11.1.1.

The moment distribution in the service state in the corner-supported element in Fig. 11.1.1 is different from that in a regular flat slab. The support moment can be expected to be more concentrated towards the support, whereas the negative moment at the other end of the support line is rather small. More reinforcement should be concentrated in the vicinity of the supported corner. It is recommended that a design support moment distribution should be used, for example, according to Fig. 11.1.2, where m_{av} is the average support moment. The length of the reinforcing bars into element *2* in Fig. 11.1.1 may be taken to be the same as into element *3*, which are determined from the rules in Section 2.10.2.

213

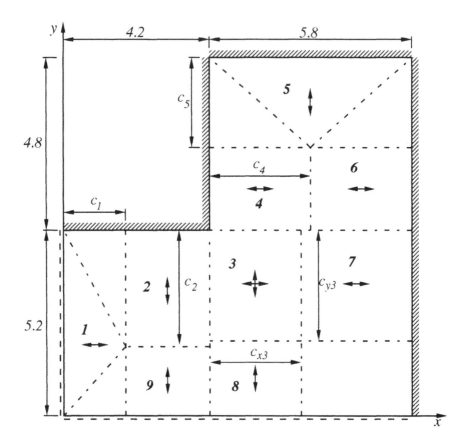

Fig. 11.1.1

The support moment in the corner-supported element under service conditions is smaller than that corresponding to a fixed edge, as it is only fixed at the corner, and may rotate around the support line further away from the corner. The support moment should thus be chosen lower than that corresponding to a fixed edge. In the examples the support moment has been taken as 75% of the moment corresponding to a fixed edge, which has been estimated to be a suitable value.

In the examples the theoretically calculated support moments have been used. As always it is permissible to increase or decrease these values, e.g. in order to limit the crack widths in the top or bottom side of the slab or to make use of minimum reinforcement.

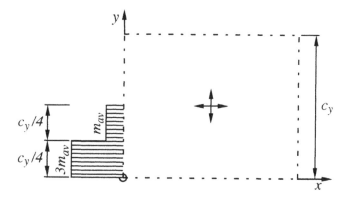

Fig. 11.1.2

The triangular end elements *1* and *5* in Fig. 11.1.1 have no direct counterpart to take the span moments, but in practice these moments can be taken care of in the slab, and the design can be based on the equilibrium of these elements, calculated from Eq. (2.4). Some particular problems related to this type of element are discussed in Example 11.4.

The design moments in the triangular corner elements in parallel with elements *2, 9, 4* and *6* in Fig. 11.1.1 may be taken as one-third of the moments in the latter elements, cf. Section 8.1.6. This also holds for the corner between elements *7* and *8*.

11.2 Reentrant corner

Example 11.1

Fig. 11.1.1 shows a slab with a reentrant corner and the elements used for its design. The slab has four fixed and two simply supported edges. The load is 9 kN/m².

The support moments at the fixed edges are calculated by means of standard formulas according to the theory of elasticity. The support moments for the corner-supported element *3* is reduced as described above. As strip *3-7* is not fully fixed at the left end and also may have a deflection at that end, the support moment for element *7* is intermediate between that for strips with the opposite ends fixed or simply supported. The *c*-values for elements *2* and *3* are calculated from Eq. (2.34) and the corresponding span moments from Eq. (2.35). The *c*-values for the triangular elements *1* and *5* are estimated using to the same principles as for rectangular slabs. Thus $c_1 = 1.8$ and $c_5 = 2.5$ are estimated to be satisfactory values. Because of symmetry $c_4 = 2.9$.

215

$$m_{f1} = \frac{9 \times 1.8^2}{6} = 4.86 \tag{11.1}$$

$$m_{s2} = -\frac{9 \times 5.2^2}{8} = -30.42 \tag{11.2}$$

$$c_2 = \frac{5.2}{2} + \frac{30.42}{9 \times 5.2} = 3.25 \tag{11.3}$$

$$m_{f2} = \frac{9 \times 3.25^2}{2} - 30.42 = 17.11 \tag{11.4}$$

$$m_{x\,s3} = -0.75 \times \frac{9 \times 5.8^2}{12} = -18.92 \tag{11.5}$$

$$m_{s7} = -\frac{9 \times 5.8^2}{9} = -33.64 \tag{11.6}$$

$$c_{x3} = \frac{5.8}{2} + \frac{18.92 - 33.64}{9 \times 5.8} = 2.62 \tag{11.7}$$

$$m_{xf3} = \frac{9 \times 2.62^2}{2} - 18.92 = 11.97 \tag{11.8}$$

$$m_{y\,s3} = -0.75 \times \frac{9 \times 5.2^2}{8} = -22.82 \tag{11.9}$$

$$c_{y3} = \frac{5.2}{2} + \frac{22.82}{9 \times 5.2} = 3.09 \tag{11.10}$$

$$m_{y f3} = \frac{9 \times 3.09^2}{2} - 22.82 = 20.15 \tag{11.11}$$

$$m_{s4} = m_{s6} = -\frac{9 \times 5.8^2}{12} = -25.23 \tag{11.12}$$

$$m_{f4} = \frac{9 \times 5.8^2}{24} = 12.62 \tag{11.13}$$

$$m_{f5} - m_{s5} = \frac{9 \times 2.5^2}{6} = 9.38 \qquad (11.14)$$

We may choose $m_{f5} = 2.38$, $m_{s5} = -7.0$. The great value of the ratio is chosen because the element is an end element in a long strip.

Fig. 11.2.1 shows the distribution of design moments. Some of the moments are so small that rules for minimum reinforcement may give higher values. If this is the case the values of c_1 and c_5 may be increased in order to make use of the reinforcement.

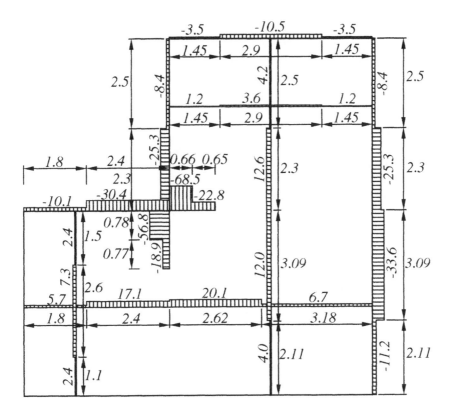

Fig. 11.2.1

The support reinforcement in elements *2* and *4* may be more concentrated towards the corner, where the risk of cracking is highest. Theoretically, these elements may be treated as corner-supported at the corner, which gives such a concentration of reinforcement. In the next example such a concentration has been chosen.

With regard to checking permissible moment distributions in the corner-supported element see Example 11.3 below. The conditions are met quite satisfactorily.

11.3 Supporting wall with a large opening

Where a supporting wall has a large opening the situation is similar to that at a reentrant corner regarding design moments for reinforcement in the direction of the wall. For the reinforcement at right angles to that direction the situation for the corner-supported element is more similar to that in a column-supported slab.

11.3.1 Inner wall

Example 11.2

Fig. 11.3.1 shows a slab with an inner supporting wall that extends to less than half the slab width. The load is 9 kN/m^2. The elements are also shown.

Elements *5, 1, 2, 3, 4* and *8* form a continuous strip. The average moments in this strip may be calculated in the same way as for a flat slab:

$$m_{s5} = -\frac{9 \times 5.5^2}{12} = -22.69 \tag{11.15}$$

$$m_{xA} = -\frac{1}{2}(\frac{9 \times 5.5^2}{12} + \frac{9 \times 5.0^2}{8}) = -25.41 \tag{11.16}$$

$$c_{x1} = \frac{5.5}{2} + \frac{25.41 - 22.69}{9 \times 5.5} = 2.80 \tag{11.17}$$

$$m_{xf1} = \frac{9 \times 2.80^2}{2} - 25.41 = 9.87 \tag{11.18}$$

$$c_{x3} = \frac{5.0}{2} + \frac{25.41}{9 \times 5.0} = 3.06 \tag{11.19}$$

$$m_{xf3} = \frac{9 \times 3.06^2}{2} - 25.41 = 16.73 \tag{11.20}$$

218

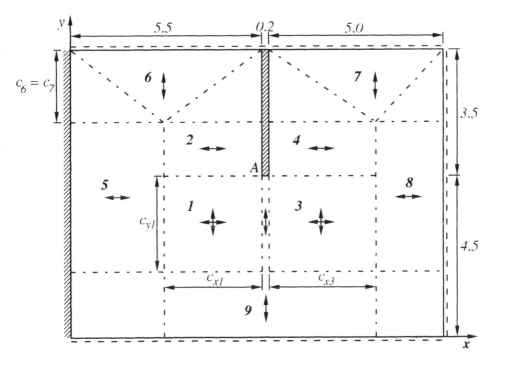

Fig. 11.3.1

Elements *1*, *3* (with intermediate one-way strip) and *9* form a strip. With the recommendation given in Section 11.1 the average moments in the strip can be calculated:

$$m_{y\,A} = -0.75 \times \frac{9 \times 4.5^2}{8} = -17.09 \qquad (11.21)$$

$$c_{y1} = \frac{4.5}{2} + \frac{17.09}{9 \times 4.5} = 2.67 \qquad (11.22)$$

$$m_{y\,f1} = \frac{9 \times 2.67^2}{2} - 17.09 = 14.99 \qquad (11.23)$$

219

If we choose $c_6 = 2.0$ m we get

$$m_{f6} = m_{f7} = \frac{9 \times 2.0^2}{6} = 6.00 \qquad (11.24)$$

Fig. 11.3.2 shows a possible distribution of the design moments. The distribution of m_{yA} follows the recommendation in Section 11.1 with a slight modification for the one-way element between the corner-supported elements. For the distribution of m_{xA} first $2m_{xA}$ has been distributed on half the width of the corner-supported elements. The same moment intensity has also been used on the first 0.6 m over the wall and $m_{xA}/3$ over the remaining part of the wall. Such a distribution can be said to correspond to an assumption that elements 2 and 4 act as corner-supported. It may also be better from the point of view of cracking than a design moment m_{xA} on the whole width (1.5 m) of elements 2 and 4.

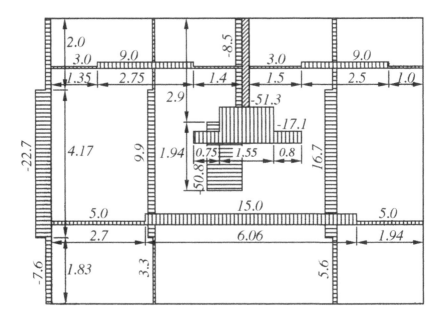

Fig. 11.3.2

11.3.2 Wall along an edge

Example 11.3

The slab in Fig. 11.3.3 has three fixed edges and one edge which is partly supported by a wall. The slab is assumed to be freely supported at the wall. If it had not been assumed to be freely supported elements *2* and *3* would have had different lengths in the y-direction and element *5* would have had to be divided into two elements. The load on the slab is 9 kN/m².

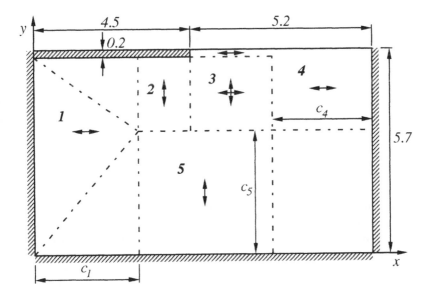

Fig. 11.3.3

A suitable value of c_I is estimated to be 3.0 m. This gives

$$m_{f1} - m_{s1} = \frac{9 \times 3.0^2}{6} = 13.50 \qquad (11.25)$$

Because this is an end element in a long slab the ratio between support and span moments is chosen to have a high value. Suitable values are $m_{f1} = 3.50$, $m_{s1} = -10.00$.

221

According to the recommendations above we get

$$m_{xs3} = -0.75 \times \frac{9 \times 5.2^2}{12} = -15.21 \tag{11.26}$$

$$m_{s4} = -\frac{9 \times 5.2^2}{9} = -27.04 \tag{11.27}$$

$$c_4 = \frac{5.2}{2} + \frac{27.04 - 15.21}{9 \times 5.2} = 2.85 \tag{11.28}$$

$$m_{f4} = \frac{9 \times 2.85^2}{2} - 27.04 = 9.51 \tag{11.29}$$

$$m_{s5} = -\frac{9 \times 5.5^2}{8} = -34.03 \tag{11.30}$$

$$c_5 = \frac{5.5}{2} + \frac{34.03}{9 \times 5.5} = 3.44 \tag{11.31}$$

$$m_{f5} = \frac{9 \times 3.44^2}{2} - 34.03 = 19.22 \tag{11.32}$$

Based on these values a distribution of support moments has been proposed in Fig. 11.3.4. For the support moment m_{xs3} a distribution with a ratio of 2 between the higher and the lower moment has been chosen, which means a little less concentration than in the previous example. The difference between these distributions probably has no practical effect.

For the corner-supported element 3, a check should be made that the conditions in Section 2.5.2 regarding moment distributions are met. In the x-direction two-thirds of the support moment are taken with $\beta=0.5$ and one-third with $\beta=0.25$. Taking a weighted average, this means that the lower limitation is $\alpha \geq (2 \times 0.25 + 0.375)/3 = 0.29$. With the distribution shown we get $\alpha = 9.51/(9.51 + 15.21) = 0.38$. The condition is met quite satisfactorily. We might even decrease the span moment somewhat (to 7.2) and put a band of reinforcement along the free edge, which may be advantageous for crack limitation.

In the y-direction we have no support moment in element 3 and we thus have to fulfil the limitations with the distribution of span moment. With the distribution shown we have $\alpha = 0.67$ for $\beta = 0.5$, which is within the limits. Also the span moment in element 2 has been given a corresponding distribution.

The distribution of support moments along the right hand-edge has been given in accordance with the general rules. This distribution is, however, not in a good agreement with the expected moments according to the theory of elasticity. If the edge had been freely

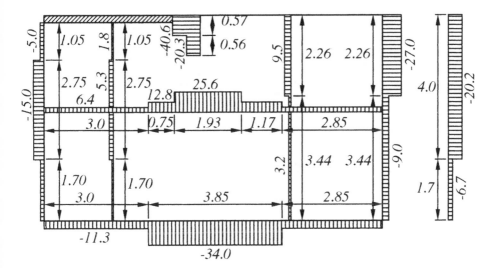

Fig. 11.3.4

supported the angle of rotation would have been greater at the centre of the edge than at the upper end, as the slab deflects more in the central part than at the end of the supporting wall. Thus the support moment for the fixed edge is greater at the centre than at the upper end, which is opposite to the moment distribution shown. For that reason it might be better to make a redistribution of the design moments, e.g. as shown to the right in the figure, particularly if cracks at the support are to be limited.

Example 11.4

The slab in Fig. 11.3.5 is the same as in the previous example with the exception that the wall opening has been moved 1.0 m to the left and a length of wall has been introduced at the right-hand end which is only 1.0 m long. Such a short wall length beside an opening introduces the problem that the triangular element *6* has to have such a small height c_6 that the moments in that element are unrealistically small. It is necessary to find a procedure for dealing with this case, which forms a continuous link between the cases corresponding to the left-hand part and the right-hand part in the previous example, i.e. with a long wall and no wall respectively. The following procedure is recommended.

The load *q* on element *4* is divided into two parts. One part *kq* is carried only in one direction, just like the whole load in the previous example. The strip carrying this load is supported on the right-hand support and has a span $l+c_6 = 6.2$ m. The other part *(1–k)q* is

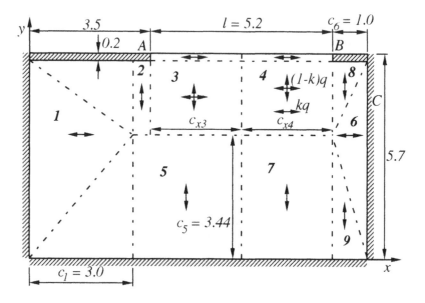

Fig. 11.3.5

carried on element *4* acting as a corner-supported element. The strip carrying this load is
supported on the end of the wall and has a span $l = 5.2$ m.

A suitable value of k is

$$k = (1 - \frac{c}{c_0})^2 \qquad (11.33)$$

where c is the length of the wall part (corresponding to the possible height c_6 of the triangular element *6*) and c_0 is the height of the triangular element, which would have been chosen if the wall part had been long. In this case the value of c_0 corresponds to $c_1 = 3.0$ m,
which gives $k = 0.44$. This is of course not an exact value, but it will be used for the analysis.

As an approximation, the support moments are given the following values:

$$m_{x\,s3} = -0.75 \times \frac{ql^2}{12} = -0.75 \times \frac{9 \times 5.2^2}{12} = -15.21 \qquad (11.34)$$

224

$$m_{x\,s4} = -0.75 \times \frac{(1-k)\,ql^2}{12} = -0.75 \times \frac{0.56 \times 9 \times 5.2^2}{12} = -8.52 \tag{11.35}$$

$$m_C = -\frac{kq\,(l+c)^2}{9} = -\frac{0.44 \times 9 \times 6.2^2}{9} = -16.91 \tag{11.36}$$

The reaction at support A in kN/m is

$$R_A = 0.56\left(\frac{9 \times 5.2}{2} + \frac{15.21}{5.2}\right) + 0.44\left(\frac{9 \times 6.2}{2} + \frac{15.21}{6.2}\right) - \frac{8.52}{5.2} - \frac{16.91}{6.2} \tag{11.37}$$
$$= 23.73$$

which gives

$$c_{x3} = \frac{R_A}{q} = \frac{23.73}{9} = 2.64 \tag{11.38}$$

$$m_{x\,f3} = \frac{9 \times 2.64^2}{2} - 15.21 = 16.15 \tag{11.39}$$

The support moment m_{xs4} cannot fall to zero just to the right of point B. If c_6 is small some part of that moment is transferred to C. It is proposed to assume that the following value is added to m_C to take into account that effect when $c_6 < c_4/2$:

$$\Delta m_C = \left[1 - \left(2\frac{c_6}{c_4}\right)^2\right] \times m_{xs4} = -\left[1 - \left(2 \times \frac{1.0}{2.60}\right)^2\right] \times 8.52 = -3.48 \tag{11.40}$$

We also have to take into account the moments in element 6:

$$m_{f6} - m_{s6} = \frac{9 \times 1.0^2}{6} = 1.50 \tag{11.41}$$

Just as for element *1*, about three-quarters of this value is taken as the support moment, which gives $m_{s6} = -1.10$. The total average moment at C (on the width $5.7 - 0.2 - 3.44 = 2.06$ m) is thus

$$m_c = -16.91 - 3.48 - 1.10 = -21.49 \tag{11.42}$$

The average moment in the remaining part of the right-hand support can be taken as $m_{s6} + \Delta m_C/3 = -2.26$.

The formal moment distribution at the right-hand support is very uneven according to this analysis, with -21.49 at C but only -2.26 on the width corresponding to c_5. In order to

225

limit cracking at the support the distribution of support moments has to be changed. A suitable distribution is proposed in Fig. 11.3.6.

Element 7 is cooperating with element 4 in carrying the part $(1-k)$ of the total load. This part of the load gives moments m_y which are $(1-k)$ times the moments in element 5. For the remaining part of the load these moments are as usual taken as one-third of the moments in element 5. The total moments m_y in element 7 are thus $(1-k+k/3) = (1-2k/3) = 0.707$ times the moments in element 5. The moments in element 5 are the same as in the previous example.

$$m_{s7} = -0.707 \times 34.03 = -24.06 \tag{11.43}$$

$$m_{yf7} = 0.707 \times 19.22 = 13.59 \tag{11.44}$$

The moment m_{yf7} is, according to normal rules, taken as $m_{xf3}/3$.

A suitable distribution of design moments is proposed in Fig. 11.3.6. It is based on the values above and values from the previous example, where these are unchanged.

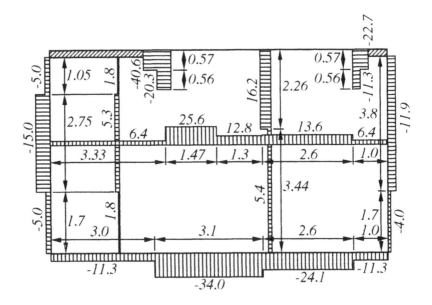

Fig. 11.3.6

11.3.3 Slab cantilevering outside wall

Example 11.5

The slab in Fig. 11.3.7 cantilevers 2.0 m outside the wall with the opening. Otherwise the slab is the same as in the previous example. The load is also the same, 9 kN/m². The cantilevering part has a fixed edge to the left in the figure, whereas the other two edges are free.

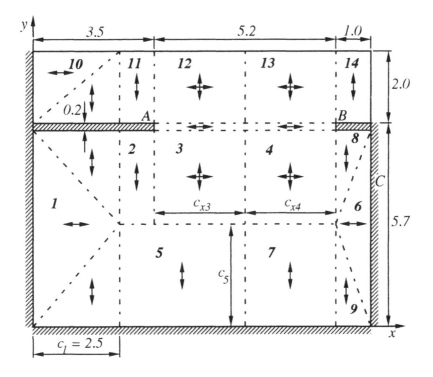

Fig. 11.3.7

The elements are also shown in the figure. The value of c_1 has been chosen somewhat lower than in the previous example because there is a support moment at the wall which decreases the deflection and also makes it less economic to carry load in the x-direction. The equilibrium equation for element *1* is

$$m_{f1} - m_{s1} = \frac{9 \times 2.5^2}{6} = 9.38 \tag{11.45}$$

A suitable choice is $m_{f1} = 2.50$, $m_{s1} = -6.88$.

As c_1 is assumed to correspond to c_0 in Eq. (11.33) we get a different value of k from that of the previous example:

$$k = (1 - \frac{1.0}{2.5})^2 = 0.36 \tag{11.46}$$

$$m_{xs3} = -0.75 \times \frac{9 \times 5.2^2}{12} = -15.21 \tag{11.47}$$

$$m_{xs4} = -0.75 \times \frac{0.64 \times 9 \times 5.2^2}{12} = -9.73 \tag{11.48}$$

$$m_C = -\frac{0.36 \times 9 \times 6.2^2}{9} = -13.84 \tag{11.49}$$

$$R_A = 0.64 \left(\frac{9 \times 5.2}{2} + \frac{15.21}{5.2}\right) + 0.36 \left(\frac{9 \times 6.2}{2} + \frac{15.21}{6.2}\right) - \frac{9.73}{5.2} - \frac{13.84}{6.2}$$
$$= 23.67 \tag{11.50}$$

$$c_{x3} = \frac{23.67}{9} = 2.63 \tag{11.51}$$

$$m_{xf3} = \frac{9 \times 2.63^2}{2} - 15.21 = 15.92 \tag{11.52}$$

$$\Delta m_C = -\left[1 - (2 \times \frac{1.0}{2.60})^2\right] \times 9.73 = -3.97 \tag{11.53}$$

As in the previous example, we have $m_{s6} = -1.10$. Adding the different moments at C we get

$$m_C = -13.84 - 3.97 - 1.10 = -18.91 \tag{11.54}$$

The average moment in the remaining part of the right-hand support can be taken as $-1.10 - 3.97/3 = -2.42$.

In the cantilevering part the moment in element *10* is the same as in *1*. For elements *12* and *13* the support moments are calculated just as in the previous examples:

$$m_{xs12} = m_{xs13} = -0.75 \times \frac{9 \times 5.2^2}{12} = -15.21 \tag{11.55}$$

Because of symmetry the c_x-values for elements *12* and *13* are 2.6 m, which is the same value as c_{x3} from the previous example. The span moment is the same as in the previous example, $m_{yf12} = 15.24$.

The cantilevering slab gives an average support moment, valid for elements *11 - 14*:

$$m_{ys} = -\frac{9 \times 2.0^2}{2} = -18.00 \tag{11.56}$$

According to the theory of elasticity, this support moment causes a change of +9.00 of the support moment for element *5*, which thus becomes (see Examples 11.3 and 11.4)

$$m_{x5} = -34.03 + 9.00 = -25.03 \tag{11.57}$$

This gives

$$c_5 = \frac{5.5}{2} + \frac{25.03 - 18.00}{9 \times 5.5} = 2.89 \tag{11.58}$$

$$m_{f5} = \frac{9 \times 2.89^2}{2} - 25.03 = 12.55 \tag{11.59}$$

As in the previous example, the moments in element *7* are taken as the moments in element *5* multiplied by $(1-2k/3)$, which in this case is 0.76:

$$m_{s7} = -0.76 \times 25.03 = -19.02 \tag{11.60}$$

$$m_{f7} = 0.76 \times 12.55 = 19.54 \tag{11.61}$$

A possible moment distribution is shown in Fig. 11.3.8. For clarity the distribution of moments over the supporting wall and its opening has been shown separated.

It should particularly be noted that there is no span moment m_y in the corner-supported elements *12* and *13*. This means that, according to the rules in Section 2.5.2, some of the support moment has to be distributed on the whole width of the element. In this case much of the moment has been concentrated near the end of the walls, corresponding to $\beta = 1.0/2.6 = 0.38$, which requires $\alpha > 0.31$. This condition is met quite satisfactorily.

As always, it is best to have more rather the less reinforcement along the long free edge. This has been taken into account in the proposed distribution.

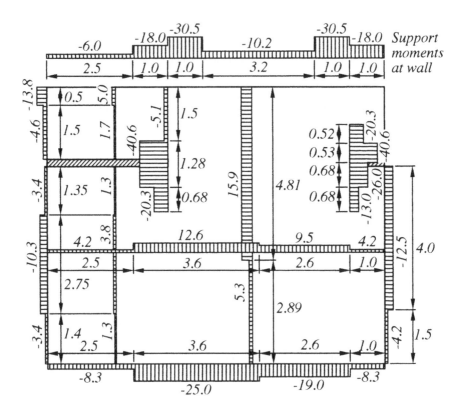

Fig. 11.3.8

CHAPTER 12

Openings in slabs

12.1 General

How the design of a slab has to be modified because of an opening depends on the shape, size and position of the opening. The most significant question is how much the opening is estimated to change the static behaviour of the slab, in particular the static behaviour in the vicinity of the opening.

Where the static behaviour of the slab is only slightly changed by the opening, the design may be based on the analysis of the slab without an opening. The reinforcement which would be cut by the opening must be arranged along its edges and properly anchored.

It is not possible to establish any simple well-founded rules for when this approximate approach is applicable. The following simple rule is recommended:

The approximate approach may be used if the opening can be inscribed in a square with side equal to 0.2 times the smallest span in the slab. It may, however, not be used for corner-supported elements, see Section 12.5, or for openings close to a free edge or in areas where torsional moments play an important role.

It cannot be proved by means of the strip method that this rule always leads to safe results. On the other hand, it is probably never possible to prove by means of yield line theory that it may lead to unsafe results. From a practical point of view it seems that the rule is satisfactory. It is also in agreement with the accepted practice of allowing small holes without any strength analysis.

Fig. 12.1.1 a) shows a simply supported slab with a uniform load and the elements which were used for its design. Fig. 12.1.1 b) shows a limiting opening for application of the recommendation above.

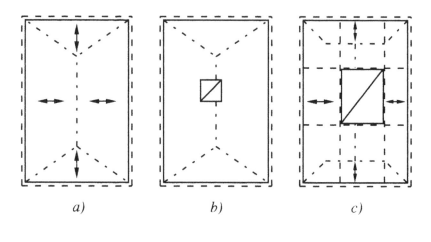

a) *b)* *c)*

Fig. 12.1.1

Fig. 12.1.1 c) shows a larger opening with support bands introduced along its edges. The lines of zero shear force have been rearranged in order to simplify the numerical analysis. Some strips are supported at one end by a support band. The support bands are either assumed to be supported at the edges of the slab or, alternatively, to be supported by support bands in the perpendicular direction. For the slab in the figure it is reasonable to assume that the support bands in the short span direction are supported at the edges of the slab, whereas the support bands in the long span direction may be partly or completely supported by the support bands in the short direction. Often a mixture of the two possibilities may be used in order to get a suitable relation between the amounts of reinforcement in the two directions.

In the analyses the support bands are assumed to have zero width and to be situated exactly along the edges of the opening (or touching the opening where they are not parallel to an edge). In practice, the design moments in the support band are distributed over a certain width within a reasonable distance from the edge, e.g. within one-third of the distance to the nearest parallel slab support. Although such a distribution is not in strict agreement with the basis of the strip method, it can be regarded as safe. It is not possible to show by means of yield line theory that it is unsafe.

In Fig. 12.1.2 the slab has fixed edges. In a) the slab is without any opening, in b) with an opening which cuts much of the main span reinforcement. As the edges of the slab are fixed, the load can be carried on cantilevers from the edges. As the edges of the opening deflect, this deflection is mainly prevented by the cantilevering action. In this case the static performance is thus changed so that certain support moments increase and take care of most of the load, whereas much of the span reinforcement can be omitted. In the first place much of the span reinforcement in the short direction may be omitted, but some span reinforcement is needed along the lower edge of the opening in the figure. A support band is also shown along this edge, which takes care of the reaction from the triangular element on the width of the opening.

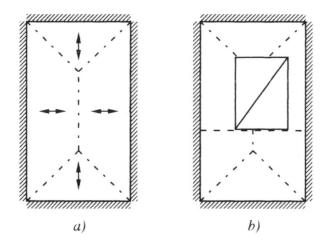

a)   *b)*

Fig. 12.1.2

The openings in Figs 12.1.1 and 12.1.2 have their edges parallel to the main reinforcement directions. If this is not the case a choice must be made between support bands in the main reinforcement directions or parallel to the edges of the opening. As a rule, the first possibility is recommended, as it gives the simplest calculations and execution.

Independent of the result of the analysis, there should always be at least one bottom and one top bar along each edge of an opening in order to decrease cracking, particularly at the corners.

Openings within the column band of a corner-supported element pose particular problems due to the high torsional moments within this area. This is treated in Section 12.5. Openings at free edges are treated in Sections 12.3 and 12.4.

12.2 Slabs with all edges supported

12.2.1 Rectangular slabs

Example 12.1

The slab in Fig. 12.2.1 is the same as in Example 3.1, but with a rather large opening *ABCD*. As in Example 3.1 the load is 9 kN/m². There is no line load along the edges of the opening.

The chosen lines of zero shear force are shown in the figure, as well as the assumed support bands along the edges of the opening.

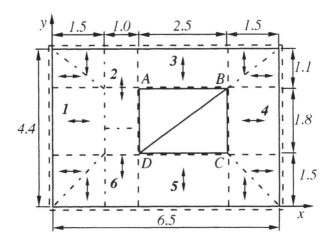

Fig. 12.2.1

The support reactions from strips *1, 3, 4* and *5* on the support bands are found by simple statics to be

$$R_{AD} = 4.05 \text{ kN/m} \tag{12.1}$$

$$R_{AB} = 4.95 \tag{12.2}$$

$$R_{BC} = 6.75 \tag{12.3}$$

$$R_{CD} = 6.75 \tag{12.4}$$

The support bands in the x-direction can be assumed to be supported either on the support bands at right angles or at their ends on the slab supports. A combination of these possibilities may also be assumed. With the proportions in the actual case it seems most suitable to assume that all or most of the load is carried by the support bands at right angles. It may well be assumed that all the load is carried in this way, but in order to demonstrate the more general case it will be assumed that 10% of the load is carried to the slab supports.

The static analysis of the support bands acted upon by these loads is trivial and only the resulting moments will be given. The first term is the maximum moment from the load carried by the support bands at right angles (span 2.5 m) and the second term is the maximum moment from the load carried to the slab supports. These two maxima are not situated at exactly the same point, but it is always safe to add them.

$$M_{AB} = 3.48 + 1.59 = 5.07 \tag{12.5}$$

$$M_{CD} = 4.75 + 2.16 = 6.91 \tag{12.6}$$

The support bands in the y-direction are loaded by the loads R_{AD} and R_{BC} respectively, by 0.9×4.95×2.5/2 = 5.57 kN at A and B, and by 0.9×6.75×2.5/2 = 7.59 kN at C and D. The resulting moments are

$$M_{AD} = 6.32 + 9.59 = 15.91 \tag{12.7}$$

$$M_{BC} = 10.55 + 9.59 = 20.14 \tag{12.8}$$

The moments in the different one-way elements are

$$m_1 = 4.96 \tag{12.9}$$

$$m_2 = 21.78 \tag{12.10}$$

$$m_3 = 1.36 \tag{12.11}$$

$$m_4 = m_5 = 2.53 \tag{12.12}$$

The triangular corner elements with $c = 1.5$ have average moments 3.38.

With values of the widths of the support bands chosen partly with respect to the values of the moments, a distribution of design moments is shown in Fig. 12.2.2.

Example 12.2

The slab in Fig. 12.2.3 is the same as in Example 3.2, but with an opening. It is also the same as in the previous example, but with the upper and left-hand edges in the figure fixed.

The fact that the edges are fixed changes the static behaviour of the parts of the slab next to these edges, as much of the load is carried by cantilever action. For this reason, no support

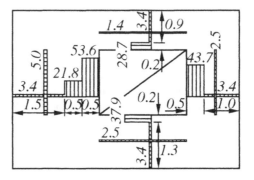

Fig.12.2.2

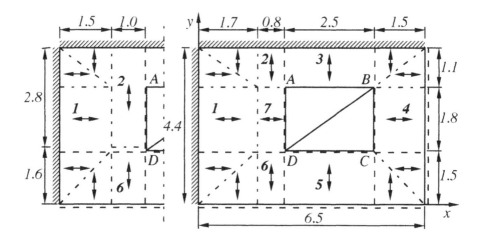

Fig. 12.2.3

band is needed along *AB*, as all the load on element *3* can be carried satisfactorily by this cantilever action. This gives a support moment

$$m_{s3} = -9 \times 1.1^2/2 = -5.45 \tag{12.13}$$

The moments in elements *4* and *5* are the same as in the previous example. This also holds for their support reactions on the support bands, $R_{BC} = R_{CD} = 6.75$ kN/m.

For the left-hand part of the slab, two possibilities will be discussed. The first possibility is shown in the main figure. The loads on elements *1* and *7* are carried in the x-direction, giving

$$m_{xf} = 9 \times 0.8^2/2 = 2.88 \qquad (12.14)$$

$$m_{s1} = 2.88 - 9 \times 1.7^2/2 = -10.13 \qquad (12.15)$$

Element *2* may be treated like element *3*, giving the same support moment and no span moment. Element *6* may, as a safe approximation, be assumed to have a span moment

$$m_{f6} = 9 \times 1.5^2/2 = 10.13 \qquad (12.16)$$

Element *7* gives a support reaction $9 \times 0.8 = 7.2$ kN/m on support band *AD*. On the same assumption as in the previous example that 90% of the load on support band *CD* is carried by the support bands at right angles, these have to carry $0.9 \times 6.75 \times 2.5/2 = 7.59$ kN at the crossing points, and the moment in band *CD* is the same as in the previous example, $M_{CD} = 6.91$ kNm, provided that the band is treated as if it were hinged at its ends.

The support bands in the y-direction are assumed to have support moments at the fixed edge of the slab. These support moments may be chosen to have the value

$$M_{sAD} = M_{sBC} = -20 \qquad (12.17)$$

Using this value we find

$$M_{fAD} = 9.65 \qquad (12.18)$$

$$M_{fBC} = 9.04 \qquad (12.19)$$

and thus a suitable ratio between support and span moments.

An alternative solution for the left hand-part of the slab is shown in the left-hand part of Fig. 12.2.3. In that case element *1* is assumed to carry all the load as a cantilever, giving

$$m_{s1} = -9 \times 1.5^2/2 = -10.13 \qquad (12.20)$$

which happens to be the same value as with the first solution. From the strip formed by elements *2* and *6* we get

$$m_{f6} = 9 \times 1.6^2/2 = 11.52 \qquad (12.21)$$

$$m_{s2} = 11.52 - 9 \times 2.8^2/2 = -23.76 \qquad (12.22)$$

As element *1* now does not give any reaction on support band *AD* this is only acted upon by the reaction force 7.59 kN from support band *CD*. Choosing

$$M_{sAD} = -8 \qquad (12.23)$$

we find

$$M_{fAD} = 4.78 \qquad (12.24)$$

which may be accepted as a suitable ratio between support and span moment with regard to the fact that the load is acting at some distance from the support.

In comparing the two solutions, we find that the results are rather similar. For the reinforcement in the x-direction the support moment is the same. There is a small span moment according to the first solution but no span moment according to the second. With due regard to minimum reinforcement requirements, this difference is, in practice, negligible. For the reinforcement in the y-direction we can compare the sum of design moments in the strip *2-6* and the support band *AD*. We then find the numerical sum of support moments to be $-0.8 \times 5.45 - 20 = -24.36$ kNm in the first case and $-1.0 \times 23.76 - 8 = -31.76$ in the second case. For span moments we find $0.8 \times 10.13 + 9.65 = 17.75$ in the first case $1.0 \times 11.52 + 4.78 = 16.30$ in the second case. These differences are unimportant. The small increase in total reinforcement in the second case is compensated for by the lack of span reinforcement in the x-direction. Either of the two approaches may be used without any noticeable difference in the behaviour of the slab.

Fig. 12.2.4 shows a distribution of design moments based on the first solution. The support moment in a support band may be distributed over a greater width than the span moment, as the support acts as a moment distributor. This has been taken into account in the figure. As usual, it is recommended that at least one bottom and one top bar are provided along each side of the opening.

Example 12.3

The slab in Fig. 12.2.5 is the same as in Example 12.1, but with a triangular opening *ABC* instead of a rectangular one. The load is still 9 kN/m^2. The same support bands in the directions of the coordinate axes are introduced, but with support bands along the edges *AC* and *BC* as well.

The load on the triangles *ACa* and *BbC* can be assumed to be carried in either the x- or y-direction, or divided between them. In this case, the y-direction is the main load-bearing direction, because of the shape of the slab. Therefore it has been chosen to carry all the load on these triangles in that direction. Thus element *5* also includes these triangles and is spanning between the edge of the slab and the support bands *AC* and *BC*. Elements *1* and *4* are supported on the support bands in the y-direction.

The support reaction from element *5* on the support bands varies linearly from $9 \times 1.5/2 = 6.75$ to $9 \times 3.3/2 = 14.85$ kN per metre width in the x-direction. The average value is

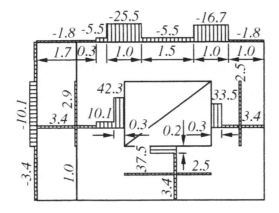

Fig. 12.2.4

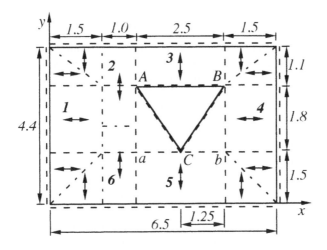

Fig. 12.2.5

10.80 kN per metre width and the difference between the ends is 8.1 kN per metre width. The support reactions from these support bands are

$$R_A = R_B = 6.75 \times 1.25/2 + 8.1 \times 1.25/3 = 7.59 \text{ kN} \tag{12.25}$$

$$R_C = 2(6.75 \times 1.25/2 + 8.1 \times 1.25/6) = 11.81 \tag{12.26}$$

R_A and R_B are carried by the support bands in the y-direction. R_C is carried by the support band ab, either only to a and b, where supported on the support bands in the y-direction, or with some part carrying to the edges of the slab. It is, as in the previous examples, assumed that 10% is carried to the edges of the slab and 90% to points a and b, and we get

$$M_C = 0.1 \times 11.81 \times 3.75 \times 2.75/6.5 + 0.9 \times 11.81 \times 2.5/4 = 8.52 \tag{12.27}$$

$$R_a = R_b = 0.9 \times 11.81/2 = 5.31 \tag{12.28}$$

Support band Aa is acted upon at A by 7.59 kN from support band AC and 5.57 kN from support band AB, see Example 12.1, at a by 5.31 kN from support band ab and between A and a by 4.05 kN/m from element 1, see Example 12.1. These loads give the moments $M_A = 17.22$, $M_a = 15.16$ and the maximum moment $M_{Aa} = 18.00$.

In the same way, support band Bb is acted upon by 13.16 kN at B, 5.31 kN at b and 6.75 kN/m between B and b. These loads give the moments $M_B = 20.14$, $M_b = 18.47$ and the maximum moment $M_{Bb} = 22.10$.

The length of support bands AC and BC is

$$l = \sqrt{1.25^2 + 1.8^2} = 2.19 \text{ m} \tag{12.29}$$

The load per metre length of these bands is equal to $1.25/l$ times the reactions per metre width in the x-direction given above. The average load is thus $10.80 \times 1.25/2.19 = 6.16$ kN/m. The moment in the support bands is, with sufficient accuracy (within 2%)

$$M_{AC} = M_{BC} = 6.16 \times 2.19^2/8 = 3.69 \tag{12.30}$$

The average moment in element 5 can be calculated by means of Eq. (2.8), taking into account the fact that the c-values in that formula correspond to half the lengths of the element, as the maximum moment is situated at the centre of the strip.

$$m_5 = \frac{9}{6 \times 4}(1.5^2 + 1.5 \times 3.3 + 3.3^2) = 6.78 \tag{12.31}$$

As the lengths of the strips forming element 5 vary considerably from the centre to the sides the design moments should be chosen with a corresponding distribution.

Fig. 12.2.6 shows a possible distribution of design moments. This figure should be compared to Fig. 12.2.2, from which some of the moments are taken.

It is interesting to compare the amount of reinforcement in the slab with an opening according to Examples 12.1 and 12.3 with that for the same slab without any opening, Example 3.1. The amount of reinforcement is proportional to the average moment over the

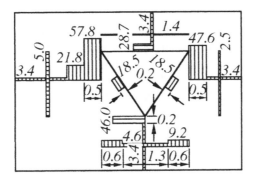

Fig. 12.2.6

whole width of the slab. In the first place, the reinforcement which is beside the opening is compared.

For the slab in Example 12.1 we find the average moments

$$m_x = (2.6 \times 3.38 + 5.07 + 6.91)/4.4 = 4.72 \tag{12.32}$$

$$m_y = (3.0 \times 3.38 + 1.0 \times 21.78 + 15.91 + 20.14)/6.5 = 10.46 \tag{12.33}$$

For the slab in Example 12.3 we find the average moments

$$m_x = (2.6 \times 3.38 + 5.07 + 8.52)/4.4 = 5.09 \tag{12.34}$$

$$m_y = (3.0 \times 3.38 + 1.0 \times 21.78 + 18.00 + 22.10)/6.5 = 11.08 \tag{12.35}$$

The average moments for the slab without any opening are, according to Example 3.1, $m_x = 6.0$, $m_y = 12.84$. A comparison shows that we would have had enough total reinforcement beside the opening if we had used the simple rule of just moving the reinforcement which would have passed through the opening to the side of it. The distribution would have been somewhat different, but this is hardly of any importance for the behaviour of the slab.

The simple rule of moving the reinforcement does not, however, tell us anything about the reinforcement in the direction between the edge of the slab and the edge of the opening (elements *1*, *3*, *4* and *5* in the examples), nor about reinforcement along the edges of the opening which are not parallel to the main reinforcement directions. For this reason the complete analysis is needed for openings above a certain size.

Example 12.4

The slab in Fig. 12.2.7 has a large circular opening. Support bands are introduced in the directions of the coordinate axes and in all 45° directions, all bands forming tangents to the circle. The analysis is similar to that in the previous example, with the 45° bands forming supports for strips in one or both directions. These bands are supported on the bands in the directions of the coordinate axes. As the analysis is similar to that in the previous example no numerical calculations are shown.

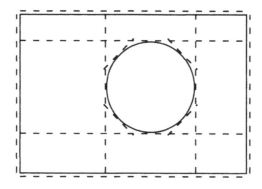

Fig. 12.2.7

As usual, it is recommended that at least one bottom bar and one top bar are provided along the edge of the opening.

12.2.2 Non-rectangular slabs

The principles for treating openings in non-rectangular slabs are the same as for rectangular slabs. The main difference is that the numerical calculations are more time-consuming because the static system and the elements are more irregular. Instead of using simple formulas it is often necessary to perform numerical summations and integrations of series of parallel strips and of support bands. Only one simple example is illustrated, a triangular slab.

Example 12.5

The triangular slab in Fig. 12.2.8 is the same as in Example 6.3, but with a large opening *ABCD*. The load is 9 kN/m².

The proposed dividing lines between elements and support bands are shown in the figure. Between elements 2 and 3 there is a line of zero shear force (maximum moment).

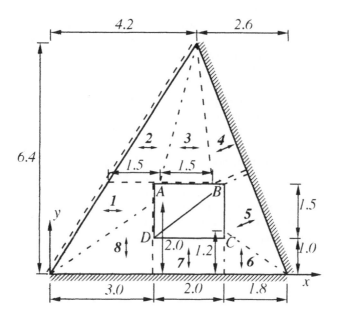

Fig. 12.2.8

Between elements *3* and *4* there is a line of zero moment. Elements *5* and *7* are assumed to function as pure cantilevers.

Element *1* is supported by the edge of the slab and the support band along *AD*. The dash-dot line in this case is just a border of the element, not a line of zero shear force. The moment in the element and the support reaction on the support band cannot be calculated by means of some simple formulas, but a numerical analysis of a series of strips in the x-direction has to be performed. The average moment in the element is 2.1 kNm/m. The support band along *AD* is assumed to be supported on the support band along *AB* and fixed at the edge of the slab. (It might also have been assumed to be extended and supported at its upper end in the figure on the support of the slab.) The numerical analysis shows that a support reaction of 4.4 kN at *A* gives a support moment of –2.94 kNm and a span moment of 1.50 kNm in support band *AD*.

The average span moment between elements *2* and *3* is $9 \times 1.5^2/6 = 3.38$ kNm/m. The reaction from element *3* on *4* is $9 \times 3.9 \times 1.5/2 = 26.33$ kN. This causes a support moment of –18.25 kNm, which is distributed on the width of the element, which is 3.92 m, giving an average moment of –4.66 kNm/m. The direct load on the element gives an average moment of –1.62, so the total average support moment in element *4* is -6.28.

The average support moment in element *5*, calculated with Eq. (2.6) is, –3.09.

The support moment in element *7* is –4.5.

Elements *6* and *8* are analysed by means of the approximation according to Fig. 2.2.4 and the resulting total moment is distributed with two-thirds as support moment and one-third as span moment. The result is $m_{s6} = -1.44$, $m_{f6} = 0.72$, $m_{s8} = -4.00$, $m_{f8} = 2.00$.

The support band along *AB* is assumed to consist of a part with positive moments, supported at its left end at the edge of the slab and at its right end on a cantilever, which is perpendicular to the fixed edge. The length of the part with positive moments is 3.0 m and the length of the cantilever is 1.04 m. The reaction 4.4 kN from support band *AD* is acting 1.36 m from the left end. The resulting span moment is 3.27 kNm and the support moment -2.07 kNm.

A distribution of design moments is proposed in Fig. 12.2.9. The span moments in the support bands are distributed over a widths of 0.5 m. The support moments from the support bands have just been added to and included in the support moments from elements *8* respectively *4* and *5*, in the latter case with about half in each. There is no reason to have a concentration of support moment at the theoretical cantilever corresponding to the support band, as the moment is small and the support acts as a distributor of the moment.

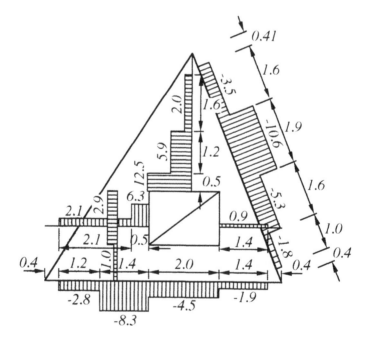

Fig. 12.2.9

The design moments close to the acute corners have been assumed to be zero, as the corresponding reinforcement would have an insufficient length within the slab and thus be useless.

As usual, there should be at least one bottom and one top bar provided along each side of the opening in order to prevent large cracks. In this case there is probably a positive moment at corner *C* under service conditions and a risk of a corner crack if there is no edge reinforcement.

The moments in this slab with an opening may be compared to those in the same slab without an opening, Example 6.3. This shows that, with the opening, a greater proportion of the load is carried in the x-direction. This seems reasonable, considering the actual size and position of the opening. In this comparison it must, however, be remembered that the ratio between loads carried in the different directions is based on a more or less arbitrary choice.

12.3 Slabs with one free edge

12.3.1 Opening not close to the free edge

This case is analysed with methods which are a mixture of those described above and the methods for slabs with one free edge described in Chapter 4.

Example 12.6

The slab in Fig. 12.3.1 is the same as in Example 4.2, but with a large opening *ABCD*. The load is 11 kN/m^2. The assumed lines of zero shear force are shown, as well as the support bands at the opening. Element *5* is assumed to carry the load as a cantilever, so no support band is needed along *DC*. The width of elements *2* and *6* is chosen so as to get suitable moments in element *1*. A support band is assumed along the free edge.

With the support reaction at the support band denoted R_{AD} we find, for element *1*,

$$m_{s1} = R_{AD} \times 3.0 - 11 \times 2.0^2 / 2 \tag{12.36}$$

With $R_{AD} = 3.0$ we find $m_{s1} = -13.0$, $m_{f1} = 3.41$, and these values seem acceptable. We further find

$$m_{f2} = 11 \times 1.3^2 / 2 = 9.30 \tag{12.37}$$

$$m_{s6} = 9.30 - 11 \times 2.5^2 / 2 = -25.08 \tag{12.38}$$

$$m_{f3} = 11 \times 1.0^2 / 8 = 1.38 \tag{12.39}$$

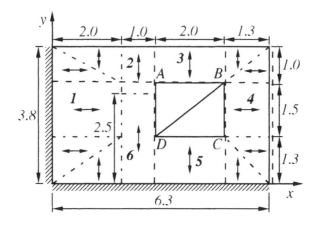

Fig. 12.3.1

$$R_{AB} = 11 \times 1.0/2 = 5.5 \tag{12.40}$$

$$m_{f4} = 11 \times 1.3^2/8 = 2.32 \tag{12.41}$$

$$R_{BC} = 11 \times 1.3/2 = 7.15 \tag{12.42}$$

$$m_{s5} = -11 \times 1.3^2/2 = -9.30 \tag{12.43}$$

The support band along *AB* can either be supported on the support bands along *AD* and *BC* or on the slab supports. In this case we will assume that it is supported on the slab supports for the whole load. This choice is made because it gives more reinforcement along the opening and more evenly distributed reinforcement in the x-direction. With the load $R_{AB} = 5.5$ acting between *A* and *B* and the reaction *R* at the right-hand end we have the following equilibrium equations

$$M_{sAB} = R \times 6.3 - 5.5 \times 2.0 \times 4.0 \tag{12.44}$$

$$M_{fAB} = R \times (1.3 + R/11) \tag{12.45}$$

With $R = 5$ we find $M_{sAB} = -12.50$, $M_{fAB} = 8.77$, which is a suitable ratio in this case. Corresponding analyses for bands *AD* and *BC* give $M_{sAD} = -3.53$, $M_{fAD} = 1.88$, $M_{sBC} = -8.69$, $M_{fBC} = 4.36$ and the reactions on the support band along the free edge equal to 1.5 kN from *AD* and 3.5 kN from *BC*.

The support band along the free edge is acted upon by $11 \times 1.3 = 14.3$ kN/m from element 2, 5.5 kN/m from element 3, triangular loads near the ends and the point loads from support bands *AD* and *BC*. Suitable moments are found to be $M_s = -42.0$, $M_f = 25.2$.

The triangular elements at the corners give average moment contributions on the relevant widths: $m_{xf} = 3.10$, $m_{xs} = -4.23$, $m_{yf} = 1.83$, $m_{ys} = -1.27$. The ratios between these moments do not follow the normal recommendations, but this is of no importance, as these moments are small compared to the other moments in the slab.

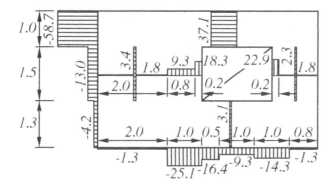

Fig. 12.3.2

Fig. 12.3.2 shows a proposed distribution of design moments. The constant m_{xf} value 37.1 on the whole width between the opening and the free edge, calculated as $8.77/1.0 + 25.2/1.0 + 3.10$, can be questioned. Maybe it would be somewhat better to concentrate the reinforcement more towards the free edge, e.g. with 49.0 on half the width and 25.2 on the other half.

The support moments from the support bands *AD* and *BC* have been distributed so that the moment is not too uneven along the edge. This type of distribution can be varied within rather wide limits without any significant influence on the behaviour of the slab. Maybe in this case it would have been slightly better to place more of the reinforcement closer to the left-hand side of the slab. The support reinforcement according to Fig. 12.3.2 is probably too small by about one metre from the left hand corner from the point of view of crack prevention.

12.3.2 Opening at the free edge

Where there is an opening at the free edge the main support band is placed just inside the opening. Beside the opening are placed support bands, which cantilever out and take care of the load closer to the free edge.

Example 12.7

The slab in Fig. 12.3.3, which has an opening at the free edge, is again basically the same as in the previous example and Example 4.2, with a load of 11 kN/m². The proposed support bands and lines of zero shear force are shown.

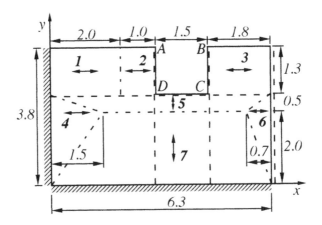

Fig. 12.3.3

In this case it is rather difficult to estimate the behaviour and moment distribution under service conditions. The opening makes the slab weaker, so that it deflects more than a slab without an opening. This larger deflection can be expected to lead to larger support moments, particularly along the short fixed edge. The position of the line of zero shear force between elements 5 and 7 determines the support moment along the long fixed edge. It has been placed some distance away from the edge in order to get support moments which are estimated to be large enough.

Applying Eqs (2.3) - (2.5) and the equation for a simply supported beam we get

$$m_{f2} = 11 \times 1.0^2/2 = 5.5 \tag{12.46}$$

$$m_{s1} = 5.5 - 11 \times 2.0^2/2 = -16.5 \tag{12.47}$$

$$m_{f3} = 11 \times 1.8^2/8 = 4.46 \tag{12.48}$$

$$m_{f6} = 11 \times 0.7^2/6 = 0.90 \tag{12.49}$$

$$m_{s4} = 0.90 - 11 \times 1.5^2/6 = -3.23 \tag{12.50}$$

$$m_{f5} = \frac{11 \times 0.5^2 (6.3 + 2 \times 4.1)}{6 \times 6.3} = 1.05 \tag{12.51}$$

$$m_{s7} = 1.05 - \frac{11 \times 2.0^2 (6.3 + 2 \times 4.1)}{6 \times 6.3} = -15.83 \tag{12.52}$$

The load from element *2* on support band *AD* is $11 \times 1.0 = 11.0$ kN/m and the load from element *3* on *BC* $11 \times 1.8/2 = 9.9$ kN/m. The corresponding design moments in the support bands are

$$M_{yD} = -11.0 \times 1.3^2/2 = -9.30 \tag{12.53}$$

$$M_{yC} = -9.9 \times 1.3^2/2 = -8.37 \tag{12.54}$$

These support bands are fixed at the lower ends in the figure. It is difficult to estimate how large the support moments are, and even the signs of the moments. It is therefore assumed that the moments are zero, corresponding to hinges. This assumption is of no importance for the behaviour of the slab, as the moments in these support bands are small compared to the other moments in the slab. With this assumption we get support reactions from these support bands on the main band

$$R_D = 11.0 \times 1.3 \times 3.15/2.5 = 18.02 \tag{12.55}$$

$$R_C = 9.9 \times 1.3 \times 3.15/2.5 = 16.22 \tag{12.56}$$

The load on the main support band along *CD* consists of these two forces and the load on element *5*. An analysis of the support band with the assumption that the point of maximum moment is 1.9 from the right-hand end gives

$$M_f = 16.22 \times 1.8 + 5.5 \times 0.7^2/3 + 5.5 \times 1.2 \times 1.3 = 38.67 \tag{12.57}$$

$$M_s = 38.67 - 18.02 \times 3.0 - 5.5 \times 1.5^2/3 - 5.5 \times 2.9 \times 2.95 = -66.57 \tag{12.58}$$

which can be considered to be a suitable moment ratio with regard to the type of load.

Fig. 12.3.4 shows a proposed distribution of design moments. The distribution along the short fixed edge has been chosen to be somewhat different from the theoretical distribution according to the analysis. A great deal of the support moment from the main support band has been moved closer to the free edge of the slab. This is motivated by the fact that it may be estimated that the moment in the service state increases towards the corner at the free edge. If the fixed edge had been freely supported the angular displacement along the edge would probably have increased towards that corner. The total sum of moments along the edge of course corresponds to the calculated values. Possibly the design moment should have been varied in more steps.

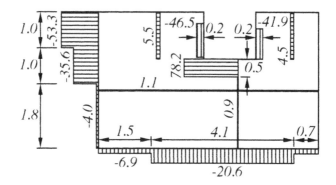

Fig. 12.3.4

12.4 Slabs with two free edges

12.4.1 Two opposite free edges

The analysis of this case is quite similar to that of a slab with one free edge, with the only difference beeing that two support bands along the free edges are introduced.

12.4.2 Two adjacent free edges and simple supports

The discussion below first considers rectangular slabs.

Where the other two edges are simply supported, no solution can be found with the simple strip method if the reinforcement is placed in the directions of the edges, see Chapter 5. This is because there have to be torsional moments to fulfil the equilibrium conditions. Such solutions are outside the scope of this book and are not treated, with the exception of some particular standard solutions, mainly for corner-supported elements.

The only possibility for treating a slab with two adjacent free edges and two simply supported edges by means of the strip method is to arrange the reinforcement in skew directions, one reinforcement direction forming a support band along the diagonal between the edges where free and simply supported edges meet. This support band acts as support for strips in a direction parallel to the other diagonal or at some other angle to the first diagonal. The reinforcement in the support band is bottom reinforcement and in the other direction it is top reinforcement. For reinforcement distributions see Section 5.2.2 and Example 7.4.

Analysis of a slab with an opening tends to become numerically complicated, and as the situation is not common, it will only be discussed in principle, without any numerical example.

If the opening is situated well outside or inside the support band the analysis is not too complicated, as the support band is unbroken. With the opening outside the support band the approach is illustrated in Fig. 12.4.1. The main support band *AC* is unbroken. Secondary support bands are introduced on both sides of the opening in the direction of the diagonal *BD*, as in the figure, or at some other angle to the main support band. The load between these bands is carried to the bands by means of bottom reinforcement parallel to *AC* or parallel to the edges of the opening. The corresponding moments can be calculated rather approximately, as they are small. The secondary support bands are supported on the main support band and on the simply supported edges.

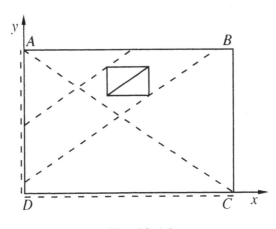

Fig. 12.4.1

If the opening is situated so that it cuts the diagonal *AC* the analysis is more complicated. One possibility is illustrated in Fig. 12.4.2. From each of the corners *A* and *C* two support bands are introduced, tangents to the opening. These support bands together take over the role of the main support band in Fig. 12.4.1. Secondary support bands are introduced past the corners of the opening. These bands are supported on the parts of the main bands between the opening and the corners *A* and *C* respectively. Each secondary band is thus supported by two main bands. The load is divided between these bands in a suitable proportion, e.g. so that the reaction of the secondary band at the simply supported edge is close to zero.

251

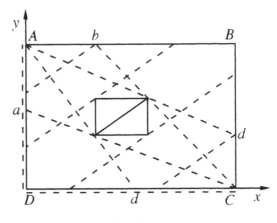

Fig.12.4.2

The band *Ad* has no real support at *d*. Another secondary band has to be introduced to take that reaction and transfer the force to the main bands. The same also holds for band *Cb* at *b*. Point *d* is not necessarily situated at the free edge. It may be assumed to be situated closer to the opening, which may give a better moment distribution.

The procedure is numerically complicated, but with a suitable choice of moment distribution it probably leads to a sound design. In addition to the designed reinforcement there should be bottom and top bars along the edges of the opening and along the free edges as well as some minimum reinforcement.

A design with the main reinforcement in the directions of the diagonals is advantageous for the behaviour of the slab and economical regarding the amount of reinforcement, as the reinforcement directions are close to the principal moment directions. However, it has the disadvantage that the reinforcing bars have varying lengths, which complicates construction work.

12.4.3 Two adjacent free edges and fixed supports

Where at least one edge is fixed, the static behaviour may be based on cantilever action, as has been shown in Chapter 5. Then it may also be possible to analyse a slab with an opening and the reinforcement parallel with the edges by means of the strip method, provided that the long edge or both edges are fixed.

Example 12.8

The slab in Fig. 12.4.3 is the same as in Example 5.1, but with an opening *ABCD*. The load is 9 kN/m^2. A formal analysis can be performed with the support bands along *AD* and *BC*, cantilevering from the fixed support, and the elements shown. The elements carrying load in the x-direction are assumed to be simply supported on the support bands.

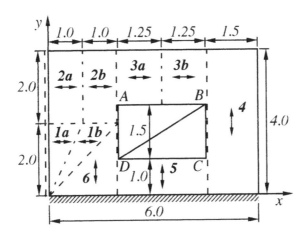

Fig.12.4.3

We get the following moments in the elements

$$m_1 = 9 \times 1.0^2/6 = 1.50 \tag{12.59}$$

$$m_2 = 9 \times 1.0^2/2 = 4.50 \tag{12.60}$$

$$m_3 = 9 \times 1.25^2/2 = 7.03 \tag{12.61}$$

$$m_4 = -9 \times 4.0^2/2 = -72.00 \tag{12.62}$$

$$m_5 = -9 \times 1.0^2/2 = -4.50 \tag{12.63}$$

$$m_6 = -\frac{9 \times 2.0^2 (2.0 + 2 \times 1.0)}{6 \times 2.0} = -12.00 \tag{12.64}$$

253

Element *2b* gives a reaction of 9×1.0 = 9.0 kN/m on support band *AD*. Elements *3a* and *3b* give reactions 9×1.25 = 11.25 kN/m on both bands. These loads give the following support moments for the bands

$$M_{sAD} = -9.00 \times 2.0 \times 3.0 - 11.25 \times 1.5 \times 3.25 = -108.84 \qquad (12.65)$$

$$M_{sBC} = -11.25 \times 1.5 \times 3.25 = -54.84 \qquad (12.66)$$

A distribution of design moments based on these values is shown in Fig. 12.4.4. The support moments in the support bands have been distributed on certain widths of the support. It has to be noted that all reinforcement in the bands has to pass beside the opening, as the moment in the bands decreases rather slowly.

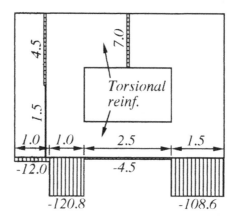

Fig. 12.4.4

The reinforcement determined from this analysis is sufficient from the point of view of safety, but it is not sufficient to make the slab behave well in the service state, particularly regarding cracks. The analysis has been based on the assumption that the load is carried by bending moments in the x- and y-directions. In reality the load in this type of slab gives rise to large torsional moments. There has to be torsional reinforcement to prevent the formation of large torsional cracks. In the analysis of the corresponding slab without an opening, Example 5.1, some of the load has been assumed to be carried by means of a solution including torsional moments, which gives at least some torsional reinforcement in the slab. No corresponding solution exists for the slab with an opening. The amount of torsional reinforcement should be at least as large in the slab with an opening as in the slab without an opening. For a realistic estimate of the torsional moments some kind of an elastic analysis has to be performed.

The distribution of support moment in Fig. 12.4.4 is also questionable. The sudden jump in value at the support bands is not realistic. In order to prevent wide cracks some extra reinforcement has to be introduced in these regions. The ratio between the support moments to the left and right of the opening is also different from what can be expected from an elastic analysis. The moment to the right of the opening is certainly larger than to the left under service conditions, as the maximum deflection of the slab is in its outer free corner. A better solution from this point of view can be obtained by supposing that the uppermost 1.5 m of element 2 is supported on support band *BC* instead of *AD*. With this change in assumption, the numerical value of M_{AD} decreases by $9.00 \times 1.5 \times 3.25 = 43.88$ to -64.96. The support reaction at band *BC* increases with 4.00 kN/m on the corresponding length, which gives an increase in the numerical value of M_{BC} of $4.00 \times 1.5 \times 3.25 = 19.50$ to -74.34. The total amount of support moment decreases and the distribution is more satisfactory with such an assumption. At the same time the span moment in the uppermost 1.5 m of element 2 increases to 10.9 kNm/m. Some part of this moment has to be added to m_3, as the force is transferred through element 3. The maximum value of the span moment within element 3 becomes 9.4 kNm/m.

It is evident that this example demonstrates a case where the strip method must be applied with caution, supplemented with at least some estimates of moment distributions in the service state, which are different from the strip method solution.

12.5 Corner-supported elements

In a corner-supported element, much of the load-bearing action takes place by means of torsional moments, which have been taken into account in the rules for the design of such elements. The importance of the torsional moments is highest close to the supported corner. An opening changes the behaviour of the slab in a way which is not easy to estimate and to compensate for. Therefore no attempt will be made to give a complete recommendation for the analysis of corner-supported elements with openings, but only some general points of view and recommendations for less crucial cases. Only rectangular elements with a distributed load will be discussed.

The type of design to be used for a corner-supported element depends on the size and position of the opening. Regarding positions at least three cases can be identified. The following recommendations are given with reference to these different situations.

Within area *A* in Fig. 12.5.1, where the two column strips cross, the torsional moments are very important. If openings of a size larger than of the order of the depth of the slab are made within this area it is recommended to refrain from the full use of the strip method and at least to supplement it by an elastic analysis. With smaller openings the reinforcement which would be cut by the opening is arranged along its edges..

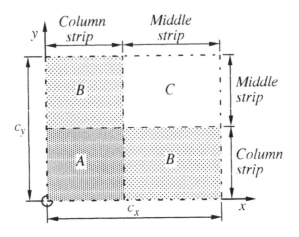

Fig. 12.5.1

Within areas *B* in Fig. 12.5.1, where a column strip and a middle strip cross, openings which can be inscribed in a square with the sides parallel to the reinforcement directions and not larger than 0.2 times the smallest *c*-value of the element may be treated in the following way:

The reinforcement which should have passed through the opening is kept in position and just cut at the opening. A corresponding amount of reinforcement is placed along the edges of the opening and given a sufficient length to ensure a safe force transfer between the two reinforcing systems.

Within area *C* in Fig. 12.5.1, where the two middle strips cross, openings which can be inscribed in a square with the sides parallel to the reinforcement direction and not larger than 0.3 times the smallest *c*-value may be treated in the same way.

The application of these recommendations will lead to slabs with adequate safety against bending failure, and probably also with acceptable behaviour under service conditions.

Where the openings are larger than those treated in these recommendations it may sometimes be possible to make an analysis based on the assumption of support bands, particularly for openings totally within area *C* in Fig. 12.5.1 or within areas *B*, but not close to the coordinate axes in the figure. Such an analysis tends to be numerically complicated and to require qualified estimates of suitable reinforcement distributions with regard to the service state. In such cases it is probably better to make an elastic analysis with a computer program.

CHAPTER 13

Systems of continuous slabs

13.1 General

Most of the previous chapters have treated one slab at a time. The only exception can be said to be flat slabs which are continuous over the column supports. In this chapter the situation will be treated where slabs are continuous over walls. The main problem is the choice of suitable support moments.

The basic approach is to start by calculating support moments for the slabs which meet over the support as if they were fixed. The support moment is then calculated as a weighted average of the moments on both sides of the supporting wall. Possibly also some of the moment difference may be taken by the wall. The weighted average is determined with regard to the estimated stiffness ratio between the slabs in accordance with normal methods applied in structural mechanics. Even if the estimate of the stiffness ratio is rather approximate the result is as a rule acceptable. It is, however, generally recommended that the support moment is chosen somewhat on the high rather than the low side.

The choice of support moment has no importance for safety but has an influence on the behaviour in the service state. Within reasonable limits it has practically no influence on the deformations, so in practice in the first place it is the risk of wide cracks that should be taken into account. Enough reinforcement must be provided in sections where it is important to avoid visible cracks. Thus the support moment must be chosen on the high side if it is mainly important to avoid wide cracks on the top surface, whereas the support moment may be chosen somewhat on the low side if it is more important to avoid wide cracks on the bottom surface. This choice may be determined by the type of floor covering, for instance.

Fig. 13.1.1 shows a situation where the support moment from the large slab is balanced not only by the support moments from the smaller slabs, but also by the moment taken by the wall at right angles, which presents a short fixed part of the support of the large slab. Close to the wall, the support moment for the large slab will be larger than if the edge had been fixed all along. In this case, the average slab moments on both sides of the wall are thus not the same, as some moment is taken by the wall at right angles.

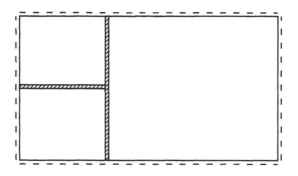

Fig. 13.1.1

Fig. 13.1.2 shows another situation where the slab moments on the two sides of the wall cannot be assumed to balance each other. The small slab has two free edges and the support moment distribution is such that it is concentrated towards the upper end in the figure, cf. Example 5.1. The support moment from the load on the large slab, which is supported all around, has its maximum closer to the centre of the support. These two moment distributions thus cannot balance each other completely, at least not in the service state. Some of the moment from the small slab has to be balanced by moments taken by the upper wall at right angles in the figure.

Fig 13.1.3 shows a slab where there is continuity only along a certain part of an edge whereas the remaining part is simply supported. In analysing this situation we may start by assuming that the whole edge is fixed. If we then release the moment from the simply supported part, the moment in the remaining part will increase. It cannot, however, increase so much that the total moment along the edge increases. With the notations in the figure, the average moment along the fixed part a_0 of the edge a must thus lie between the values m_0 and am_0/a_1, where m_0 is the moment when the whole edge is fixed. Which is the best value between these limits has to be estimated. With the proportions shown in the figure the moment will be largest at the end of the continuous part.

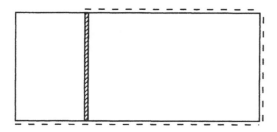

Fig. 13.1.2

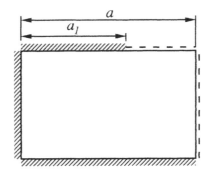

Fig. 13.1.3

13.2 Systems of rectangular slabs

In order to limit the numerical calculations, the examples chosen below are as simple as possible. The intention is only to demonstrate the principles. With a proper understanding of the general approach there should be no problem in analysing more complex slab systems.

Example 13.1

The slab system in Fig. 13.2.1 consists of only two slabs, which meet over a supporting wall. All other edges are simply supported. The load is 9 kN/m^2. Lines of zero shear force are shown, based on the general principles given in Section 2.2.

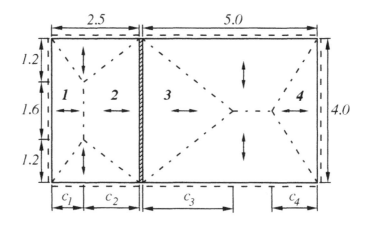

Fig.13.2.1

We start by calculating the support moments as if the slabs had fixed supports at the wall. Applying Eqs (2.5) and (2.4) and directly subtracting the span moments from the numerical values of support moments we find the average moments

$$m_{s2} = \frac{9\,(c_1^2 - c_2^2)\,(4.0 + 2 \times 1.6)}{6 \times 4.0} = 2.70\,(c_1^2 - c_2^2) \tag{13.1}$$

$$m_{s3} = \frac{9\,(c_4^2 - c_3^2)}{6} = 1.50\,(c_4^2 - c_3^2) \tag{13.2}$$

Suitable choices of c-values (with regard to ratios between span and support moments) are: $c_1 = 0.9$, $c_2 = 1.6$, $c_3 = 2.6$, $c_4 = 1.3$, which gives $m_{s2} = -4.73$, $m_{s3} = -7.61$. As the small slab is stiffer than the large slab the final moment should be chosen closer to -7.61 than to -4.73, say $m_s = -6.75$.

We now have to adjust the c-values to conform to the new value of m_s. We can assume that the dimensions in the directions at right angles are unchanged. (In more complex slab systems of course changes in these directions may also have to be taken into account.) We can then apply the above equations.

For the small slab we have to take into account that $c_1 + c_2 = 2.5$. We find $c_1 = 1.75$, $c_2 = 0.75$. For the large slab we may keep the value of c_4, as the change in the support moment in this case influences the moment in element *4* very little. We find $c_3 = 2.49$.

Example 13.2

The slab system in Fig. 13.2.2 is the same as in the previous example with the exception that the small slab has a long free edge. This of course increases the support moment. It is now more difficult to estimate the pattern of lines of zero shear force for the small slab. It seems reasonable to assume $c_1 = 0.5$ when the support is fixed, although a smaller value might be chosen. The equilibrium equation gives

$$m_{s2} = \frac{9(c_1^2 - c_2^2)(4.0 + 2 \times 1.0)}{6 \times 4.0} = 2.25\,(c_1^2 - c_2^2) = -8.44 \qquad (13.3)$$

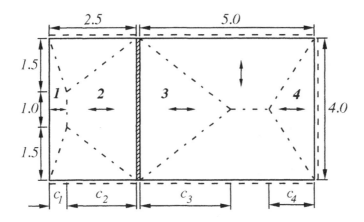

Fig. 13.2.2

For the large slab, Eq. (13.2) is valid, and for the case with a fixed support we use the same values as above, i.e. $m_{s3} = -7.61$.

Because of the free edge the small slab is now less stiff than the large slab. A suitable value of the support moment may be -8.10. Applying the same procedure as above we find $c_1 = 0.53, c_2 = 1.97, c_3 = 2.66$.

The result in this case is very sensitive to the choice of c_1. This, however, does not influence the safety, only the behaviour in the service state.

Example 13.3

The slab system in Fig. 13.2.3 is again the same, but with the exception that now the upper edge of the small slab is also free. For that slab we apply the same type of solution as in Example 5.1, starting with an assumption that the continuous edge is fixed. Most of the load is carried by the strips in the figure, but part of it is carried by a solution including torsional moments, as discussed in Chapter 5. As in Example 5.1, we assume that 20% of the load is carried in this way. If we assume $c_1 = 1.0$, the remainder, 7.2 kN/m^2, gives rise to a support moment

$$m_{s2} = -\frac{7.2 \times 2.5^2 (4.0 + 2 \times 3.0)}{6 \times 4.0} = -18.75 \tag{13.4}$$

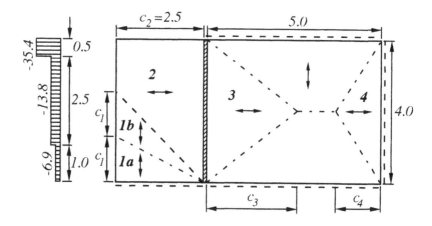

Fig. 13.2.3

This moment is unevenly distributed along the support, with the largest value in the vicinity of the upper corner, see Fig. 5.2.4. The whole of the moments m_{s2} can not be used to balance $m_{s3} = -7.61$, as their distributions are so different. One part of m_{s2} has to be carried by the perpendicular support for the large slab. We can, for example, assume that 20% is taken in that way. The remaining part, -15.0, might be taken into account in determining the value of m_s.

As the small slab is now much less stiff than the large slab, a suitable support moment might be about -13. However, this large increase in m_{s3} corresponds to an appreciable angular displacement over the support. This angular displacement will cause more of the support moment m_{s2} to be carried at the upper corner, and also more to be carried by torsional moments.

The situation is evidently very complex, and it is difficult to know how the moments and the type of static function can be best estimated. It is proposed that the part of the load taken by the solution including torsional moments should be increased to 30% and c_1 should be increased to 1.5 m. We then find $m_{s2} = -14.77$. If we further assume that 30% is carried at the upper corner we find $m_s = -10.34$. This value seems reasonable with respect to the large slab, and gives $c_3 = 2.93$.

This solution is to a rather large extent evidently based on estimates. It is, however, safe as long as the assumed moment distributions are such that the corresponding reinforcement can function in an appropriate way. The chosen moment distribution has to cover an appropriate moment distribution for the large slab as well as for the small slab. This will be the case if we distribute the support moment according to the diagram shown to the left in the figure. A normal distribution, with two-thirds of m_s outside the quarter points and four-thirds inside the quarter points for the large slab is covered, as well as the total moment for the small slab with a strong concentration towards the upper corner, where the slab is locally fixed. The solution also seems to give an acceptable reinforcement distribution with regard to the service state.

Example 13.4

The slab system in Fig. 13.2.4 consists of three slabs A, B and C, supported on walls between the slabs but simply supported along all other edges. All the slabs have a uniform load of 9 kN/m².

For slab A we can calculate the average support moment by assuming $c_1 = 1.8, c_3 = 3.2$, which gives

$$m_{s3} = \frac{9 \times (1.8^2 - 3.2^2) (6.0 + 2 \times 3.0)}{6 \times 6.0} = -21.00 \tag{13.5}$$

For slab B we can assume $c_5 = 2.5, c_7 = 1.3$, which gives the average moment

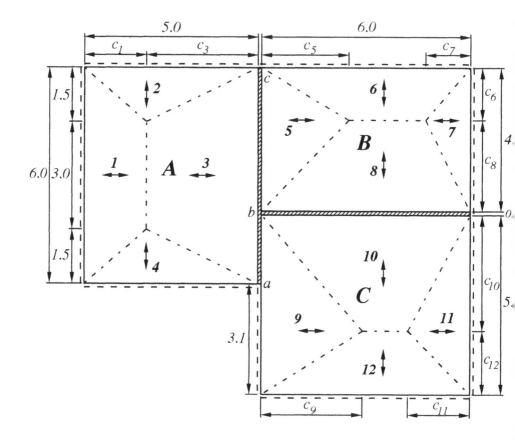

Fig. 13.2.4

$$m_{s5} = \frac{9 \times (1.3^2 - 2.5^2)}{6} = -6.84 \tag{13.6}$$

Slab C has an edge which is only partly continuous. We then have to start by assuming that it is fully fixed. With $c_9 = 3.2$, $c_{11} = 1.8$ we get

$$m_{s9} = \frac{9 \times (1.8^2 - 3.2^2)}{6} = -10.50 \tag{13.7}$$

The total moment corresponding to this average moment is $-10.50 \times 5.0 = -52.5$ kNm. From the shape of element 9 it can be seen that the greatest part of this moment is theoretically caused by strips which end at the part of the edge which is in reality simply supported. When the whole edge is fixed thus, only a minor part of the total moment falls within the distance ab. When the moment on the freely supported part is released, however, the moment on ab will increase, with a concentration close to a. It may be estimated that between 30 and 60% of the total moment will remain within ab, say, 45% or about -24 kNm. Most of this total moment will be taken close to a and transformed to the wall at right angles at a. We may assume that the moment is taken as a distributed moment $m_{s3-9} = -5$ kNm/m and the remaining part as a concentrated moment $M_{sa} = -14.5$ kNm at a.

In determining the support moments along the support abc we have to take into account that m_{s3} is the largest and will cause an angular displacement of the slab over the support. This angular displacement will be prevented at b by the wall at right angles. This causes a concentrated support moment at b. An estimate may be that the moment difference on some 20 to 30% of the support length will be taken by that concentrated moment. The moment difference is about 15 kNm/m and the concentrated moment can thus be taken to be $M_{sb} = -22$ kNm.

The remaining total moment $-21.00 \times 6.0 + 22 = -104$ is mainly taken at bc. We can assume $m_{s3-9} = -10$ and $m_{s3-5} = -21$, which gives approximately the right total moment.

Using the above moments, and averaging with an approximate regard to the stiffnesses, gives the design moments $m_{s3-9} = -8$, $m_{s3-5} = -15$.

After these moments have been determined we can calculate the new c-values. Thus with the average design support moment $m_{s3} = -(21 + 1.9 \times 8 + 4.0 \times 15)/6.0 = -16.03$ we get

$$\frac{9(c_1^2 - c_3^2)(6.0 + 2 \times 3.0)}{6 \times 6.0} = -16.03 \tag{13.8}$$

which leads to $c_1 = 1.97, c_3 = 3.03$.

In slab B we may keep c_7 unchanged and equal to 1.3, and we get

$$\frac{9(1.3^2 - c_5^2)}{6} = -15 \tag{13.9}$$

which gives $c_5 = 3.42$.

In slab C we may keep c_{11} unchanged and equal to 1.8. The average moment m_{s9} is $-(14.5 + 1.9 \times 8)/5.0 = -5.94$ and we get

$$\frac{9(1.8^2 - c_9^2)}{6} = -5.94 \tag{13.10}$$

which gives $c_9 = 2.68$.

With the changed positions of the lines of zero shear force we can now calculate the average fixed edge support moments over the wall between slabs B and C, assuming $c_6 = 1.4$, $c_8 = 2.6$, $c_{10} = 3.2$, $c_{12} = 1.8$

$$m_{s8} = \frac{9(1.4^2 - 2.6^2)(6.0 + 2 \times 1.28)}{6 \times 6.0} = -10.27 \tag{13.11}$$

$$m_{s10} = \frac{9(1.8^2 - 3.2^2)(6.0 + 2 \times 1.52)}{6 \times 6.0} = -15.82 \tag{13.12}$$

Slab B is slightly stiffer than C and a suitable design moment is -13.5.

The remaining part of the analysis follows normal procedures. In determining the distribution of design moments, the concentrated moments at a and b should be distributed on small widths and added to the other moments. In slab C the span moment from elements 9 and 11 must be rather unsymmetrically distributed in order to compensate for the unsymmetrical distribution of the support moment.

Example 13.5

A water tank has a rectangular bottom 4.0 m×6.0 m and a depth of 4.0 m. It rests on columns along the edges of the bottom. It must be designed for the case where it is filled with water. The bottom B of the tank and two of the sides A and C are shown in Fig. 13.2.5. Patterns of lines of zero shear force are also shown. Because of symmetry $c_1 = c_3$ etc.

The water pressure is assumed to increase by 10 kN/m² per metre depth (a more correct value is 10.2). It is thus 40 kN/m² at the bottom.

Assuming $c_2 = 1.6$, $c_3 = 2.4$, $c_5 = 2.0$, $c_9 = 2.0$, $c_{11} = 1.5$, and taking the symmetry into account, we get the following average moments, applying the equations in Sections 2.3.2-4:

$$m_2 = \frac{16 \times 1.6^2(6.0 + 3 \times 1.2)}{12 \times 6.0} = 5.46 \tag{13.13}$$

$$m_{s4} = 5.46 - \frac{40 \times 2.4^2(6.0 + 2 \times 1.2)}{6 \times 6.0} + \frac{24 \times 2.4^2(6.0 + 3 \times 1.2)}{12 \times 6.0} = -29.87 \tag{13.14}$$

$$m_{f3} - m_{s3} = \frac{40 \times 2.4^2(4.0 + 2 \times 1.6)}{24 \times 4.0} = 17.28 \tag{13.15}$$

$$m_{f6} - m_{s6} = \frac{40 \times 2.0^2(6.0 + 2 \times 2.0)}{6 \times 6.0} = 44.44 \tag{13.16}$$

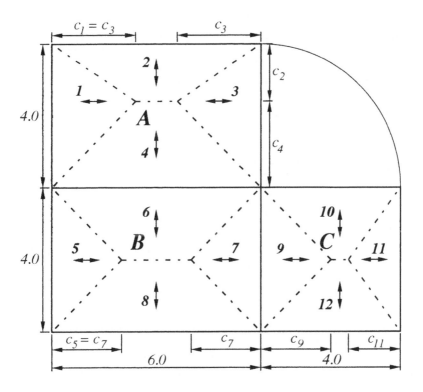

Fig. 13.2.5

$$m_{f7} - m_{s7} = \frac{40 \times 2.0^2}{6} = 26.67 \tag{13.17}$$

$$m_{11} = \frac{15 \times 1.5^2}{12} = 2.81 \tag{13.18}$$

$$m_{s9} = 2.81 - \frac{40 \times 2.0^2}{6} + \frac{20 \times 2.0^2}{12} = -17.19 \tag{13.19}$$

$$m_{f10} - m_{s10} =$$

$$\frac{40 \times 2.0^2 [3 \times 1.5^2 + 6 \times 0.5 \ (2 \times 1.5 + 0.5) + 2.0 \ (4 \times 4.0 - 3 \times 2.0)]}{24 \times 4.0^2} = 15.52 \qquad (13.20)$$

With regard to suitable moment ratios, we may choose $m_{s3} = -13$, $m_{s6} = -30$, $m_{s7} = -20$, $m_{s10} = -10.5$.

Elements *2* and *11* are supported by support bands. The moments in these support bands are

$$M_{f2} - M_{s2} = \frac{16 \times 3.0^2 \times 1.6}{4} - \frac{16 \times 2.4^2 (1.6 + 2 \times 1.6)}{24} = 39.17 \qquad (13.21)$$

$$M_{f11} - M_{s11} = \frac{15 \times 2.0^2 \times 1.5}{4} - \frac{15 \times 2.0^2 (1.5 + 2 \times 1.5)}{24} = 11.25 \qquad (13.22)$$

With regard to load distributions and suitable moment ratios we may choose $M_{s2} = -25$, $M_{s11} = -7$ as support moments when the ends are fixed.

If we compare the moments on both sides of supports we find that the moment differences are quite small in all cases except for the support bands. In the first place we therefore take this difference into account.

As slab *C* is smaller and stiffer than *A*, a suitable design support moment for the support bands, based on the above values, is –17 kNm. (Application of the recommendation in Section 10.2 gives $-(39.17+11.25)/3 \approx -17$, thus the same result.) Such a value will mean that the span moment in the band along element *11* is negative and the band is bent inwards. On the other hand, the band along element *2* will deflect outwards. These deformations will lead to a decrease in the numerical value of m_{s9} and an increase in m_{s4}. Taking this and the moment differences at the edges into account, we can try new c-values, which are now regarded as final.

We start by assuming new values for slab *C*, $c_9 = 1.9$, $c_{11} = 1.6$. This leads to $m_{11} = 3.41$, $m_{s9} = -14.94$, $m_{f10} - m_{s10} = 15.98$, $M_{f11} - M_{s11} = 12.80$.

Based on $m_{s7} = m_{s9} = -14.94$ we can find that a suitable value of c_7 is 1.75, which gives $m_{f7} = 5.48$.

If we further assume $c_2 = 1.5$, $c_4 = 2.5$ and $m_{s3} = m_{s10} = -12$, we find that a suitable value of c_3 is 2.5, which gives $m_{f3} = 6.23$.

We have now determined all c-values and we can calculate $m_2 = 4.22$, $m_{s4} = -31.80$, $M_{f2} - M_{s2} = 33.05$. A suitable value of the support moment for the support bands is –15, which gives $M_{f2} = 18.05$, $M_{f11} = -2.20$, a small negative value.

Applying $m_{s6} = m_{s4} = -31.80$ we find $m_{f6} = 17.09$, which is acceptable.

Because of the triangular load distribution and the shape of the tank, strong moment concentrations cannot be expected at the free edges. The design moment distribution might

268

therefore be chosen to be rather even. On the other hand, the analysis is based on the assumption of support bands along the free edges. The moment distribution proposed in Fig. 13.2.6 for slab *A* is a compromise between these points of view.

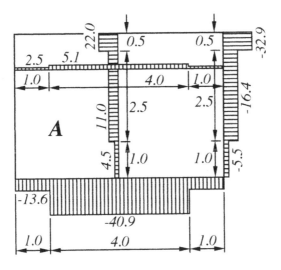

Fig. 13.2.6

The result of this analysis is a tank with adequate safety and a reinforcement distribution which is suitable from the point of view of deformations. It may not be the best reinforcement arrangement to limit cracking. This type of design is therefore not recommended where the concrete is intended to be watertight. For such a case the design should be based on the theory of elasticity, taking due account of torsional moments.

13.3 Rectangular slabs and concrete walls

Concrete walls can be regarded as concrete slabs with zero load. The support moments can therefore, as a reasonable approximation, be calculated with the same methods as above, with the support moments for fixed edges equal to zero for the walls, taking into account the ratios between the stiffnesses of the slabs.

13.4 Other cases

Systems including non-rectangular slabs or slabs with corner-supported elements can be analysed according to the same principles as above. Of course such cases tend to give more laborious numerical calculations and more intricate estimates. Therefore no numerical examples are shown.

CHAPTER 14

Joist floors

14.1 General

In a joist floor *joists* or *ribs* in one or two directions interact with a rather thin slab to take the moments. For positive moments the joists and the slab act as T-beams with tension reinforcement in the bottom of the joists and compressive stresses in the slab. Joist floors can be made to take much higher positive than negative moments. In places where large negative moments occur, the joist floor is replaced by a solid slab with the same depth.

A joist has a very limited ability to take torsional moments in the direction of the ribs. The design has to be made on the assumption that only bending moments occur. This limits the possibility of using joist floors in situations where torsional moments are essential for carrying the load. Thus, for example, a joist floor with joists parallel to the edges cannot be used for a rectangular slab with two adjacent edges free and the other edges simply supported. Where a corner-supported element forms part of a joist floor there must be a solid part in the vicinity of the supported corner.

As joist floors are designed on the assumption that they take no torsional moments, but only bending moments in the directions of the joists, the application of the strip method is based on the simple strip method.

A joist floor has a limited ability to redistribute moment in the lateral direction. The lateral redistribution within wide limits, which may be accepted for a solid slab, cannot be accepted within a joist floor. Each joist has, in principle, to take the moment from the load which has been assigned to it. Where there is a system of crossing joists some lateral moment redistribution may however be accepted.

14.2 Non-corner-supported floors

The simple strip method may in principle be applied directly. Just as for solid slabs, the floor is normally divided into areas where the load is carried in the different joist directions. The dividing lines are often lines of zero shear force and of maximum positive moments. The load which is assigned to each joist is that on the area between the centre lines between the parallel joists.

As the lines of zero shear force often form an angle to the direction of a joist, the theoretical load distribution on a joist from a uniform load on the floor has a shape according to Fig. 14.2.1. The moments in the joist are then determined from the equation

$$M_f - M_s = \frac{ql^2}{2} + \frac{q(\Delta l)^2}{24} \tag{14.1}$$

The second term can often be disregarded.

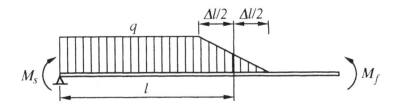

Fig. 14.2.1

<u>*Example 14.1*</u>

The simply supported joist floor in Fig. 14.2.2 has joists at 0.9 m centres in both directions. The load is 8 kN/m², which corresponds to 7.2 kN/m for the loaded parts of the joists. An assumed pattern of lines of zero shear force is shown.

Joist a has $l = 0.79$ and $\Delta l = 0.79$. We thus get

$$M_a = 7.2 \times 0.79^2/2 + 7.2 \times 0.79^2/24 = 2.25 + 0.19 = 2.44 \tag{14.2}$$

In the same way we find

$$M_b = 7.2 \times 1.58^2/2 + 0.19 = 9.18 \tag{14.3}$$

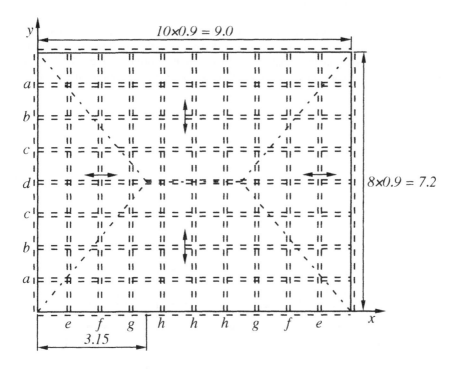

Fig. 14.2.2

$$M_c = 7.2 \times 2.36^2/2 + 0.19 = 20.24 \qquad (14.4)$$

$$M_d = 7.2 \times 2.95^2/2 + 7.2 \times 0.39^2/24 = 31.38 \qquad (14.5)$$

Notice that Δl is smaller for joist d.
Joist e has $l = 1.03$, $\Delta l = 1.03$ etc. Joist h has $\Delta l = 0$. We get

$$M_e = 7.2 \times 1.03^2/2 + 7.2 \times 1.03^2/24 = 3.82 + 0.32 = 4.14 \qquad (14.6)$$

$$M_f = 7.2 \times 2.06^2/2 + 0.32 = 15.60 \qquad (14.7)$$

$$M_g = 7.2 \times 3.09^2/2 + 0.32 = 34.69 \qquad (14.8)$$

$$M_h = 7.2 \times 3.6^2/2 = 46.66 \qquad (14.9)$$

273

A certain limited redistribution of moments between adjacent parallel strips may be performed.

This solution gives a rather uneven reinforcement distribution. It is also possible to make an analysis on the assumption that the load is divided between the two directions within certain areas. We can, for example, assume a load distribution according to Fig. 14.2.3, where half the load is taken in each direction within the corner regions. With this distribution we get

$$M_a = M_b = M_e = M_f = 3.6 \times 2.25^2/2 = 9.11 \tag{14.10}$$

$$M_c = M_d = 7.2 \times 2.25^2/2 = 18.23 \tag{14.11}$$

$$M_g = M_h = 46.66 \tag{14.12}$$

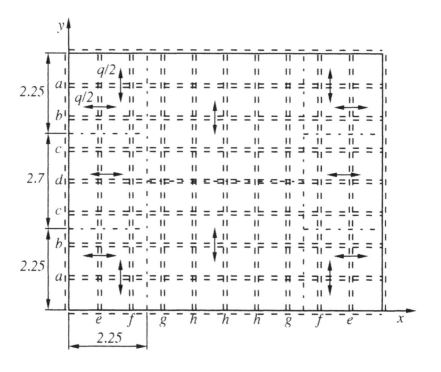

Fig. 14.2.3

This analysis is simpler and gives a more even reinforcement distribution. It gives a little more reinforcement. The sum of all moments is for the first analysis 2(2.44 + 9.18 + 20.24 + 4.14 + 15.60 + 34.69) + 31.38 + 3×46.66 = 343.94 and for the second analysis 8×9.11 + 3×18.23 + 5×46.66 = 360.87, which is about 5% higher, corresponding to 5% more reinforcement. This small difference is in practice more or less compensated for by a greater need for minimum reinforcement in the first case.

In many cases a solution of the second type is to be preferred. Of course, other load distributions than one with half in each direction may be used.

Example 14.2

The joist floor in Fig. 14.2.4 has one free edge and three fixed edges. The joist spacing is 1.2 m. Along the fixed edges there is a solid slab of width equal to one joist spacing plus half the width of a joist. Along the free edge there is a beam with a width equal to half the width of a joist plus 0.2 m. The load is assumed to be a uniform 8 kN/m². The higher dead weight of the solid parts is thus disregarded in order to simplify the numerical calculations.

A simple approach has been chosen with dividing lines parallel with the edges and half the load carried in each direction on areas 3.0 m×3.0 m.

We start by analysing joist *b*, which has a load of 1.2×8 = 9.6 kN/m at 3.0 m from each end.

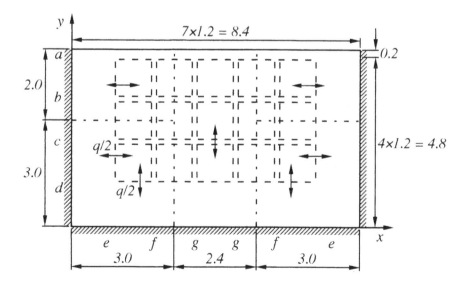

Fig.14.2.4

$$M_{fb} - M_{sb} = 9.6 \times 3.0^2/2 = 43.2 \qquad (14.13)$$

Due to the higher stiffness near the supports and to the load distribution we may choose $M_{sb} = -32.0$, $M_{fb} = 11.2$. The moments in joist c may be taken as one-half of these values, $M_{sc} = -16.0$, $M_{fc} = 5.6$.

Strip d, which is a solid slab strip, must take load from a width of 1.8 m and will thus take a moment which is 1.5 times that in joist c, $M_{sd} = -24.0$, $M_{fd} = 8.4$. The reinforcement for these moments must mainly be placed far from the parallel support.

Joist f has a load of 4.8 kN/m at 3.0 m from the fixed end and is supported on beam a, where the support reaction is denoted R_{fa}. The equilibrium equation is

$$M_{sf} = -4.8 \times 3.0^2/2 + 4.9 R_{fa} = -21.6 + 4.9 R_{fa} \qquad (14.14)$$

Suitable values may be $R_{fa} = 1.6$ kN, $M_{sf} = -13.76$, which gives the span moments $M_{ff} = 3.31$. The ratio between moments is acceptable with respect to the higher stiffness near the fixed support and the load distribution. The moments in strip e can be taken as 1.5 times these values, $M_{se} = -20.64$, $M_{fe} = 4.97$ and the corresponding reaction $R_{ea} = 2.4$ kN, acting 0.9 m from the support.

Joist g has a load of 9.6 kN/m on the whole length. The equilibrium equation is

$$M_{sg} = -9.6 \times 5.0^2/2 + 4.9 R_{ga} = -120 + 4.9 R_{ga} \qquad (14.15)$$

Suitable values may be $R_{ga} = 17$ kN, $M_{sg} = -36.7$, $M_{fg} = 13.35$.

The edge beam a has to carry a uniform load of $0.8 \times 8 = 6.4$ kN/m acting along a length of 3.0 m from each end and the reactions $R_{ea} = 2.4$, $R_{fa} = 1.6$ and $R_{ga} = 17$ kN. The equilibrium equation is

$$M_{fa} - M_{sa} = 6.4 \times 3.0^2/2 + 2.4 \times 0.9 + 1.6 \times 2.4 + 17 \times 3.6 = 96.00 \qquad (14.16)$$

Suitable values may be $M_{sa} = -64.00$, $M_{fa} = 32.00$.

The largest positive joist moment is 13.35 kNm and the largest negative joist moment where it meets the solid slab can be shown to be about -6.4 kNm. The most stressed member is the edge beam. Whether the design is acceptable can only be judged after the reinforcement has been designed and the concrete stresses calculated. It may have been advantageous to choose a wider edge beam.

14.3 Floors with corner-supported elements

In a corner-supported element the moment field in the vicinity of the support is nearly polar symmetric with negative tangential moments and positive or numerically smaller negative radial moments. This moment field contains large torsional moments with respect to the reinforcement directions. These torsional moments cannot be carried by a joist floor. In the vicinity of the support there has to be a solid slab, with the ability to take torsional moments.

A corner-supported element which forms part of a joist floor thus has to have a solid slab in the vicinity of the supported corner, whereas the rest of the element has crossing joists. As a rule, the solid part has a rectangular shape and has its edges as a continuation of the joists. Only this case will be treated here, but of course the same principles may be used even where the solid part has another shape.

A corner-supported element thus has crossing joists which carry the load only by means of bending moments. These joists are in their turn carried by the solid part which also takes torsional moments and which carries the load into the support.

Fig. 14.3.1 shows a corner-supported element which has a solid part next to the supported corner. The element is divided into column strips and middle strips. The boundary lines between column and middle strips are halfway between the solid part and the nearest joist. The width of the column strip in the x-direction is denoted $\beta_y c_y$.

Within the area where the middle strips cross the load is divided between the two strips. The simplest assumption is to divide the load into two equal parts, and only this case will be treated here. Of course it is possible to make other assumptions, but this does not seem to give any advantage in this case.

A middle strip is supported on the crossing column strip. In the analysis we may use average moments and forces per unit width. From these average moments and forces we later calculate the moments and forces in the joists and in the solid part. The forces per unit area on the middle strip in the x-direction are shown in Fig. 14.3.2. The reaction on the column strip is assumed to be uniformly distributed. For the moments per unit width in the middle strip in the x-direction we find, with index m for middle strip

$$m_{xfm} - m_{xsm} = q(1 - \beta_x)c_x^2/4 \qquad (14.17)$$

It can be demonstrated that it is suitable to have equal numerical values for the two moments, thus

$$m_{fm} = -m_{sm} = q(1 - \beta_x)c_x^2/8 \qquad (14.18)$$

The value of the support moment may have to be modified in order to make it agree with the value from the element on the other side of the support. In such a case the value of the span moment has to be modified so that Eq. (14.17) is fulfilled.

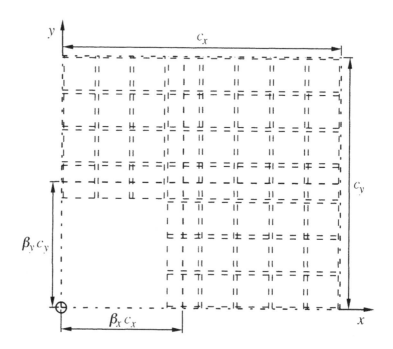

Fig. 14.3.1

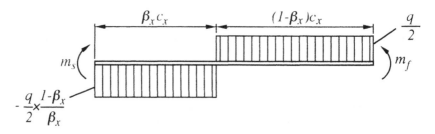

Fig. 14.3 2

The load from the middle strip in the x-direction on the column strip in the y-direction is given in Fig. 14.3.2. The column strip also has to carry the direct load q. The total load on the column strip within the joisted part is

$$q_{yc} = \frac{q(1-\beta_x)}{2\beta_x} + q = \frac{q(1+\beta_x)}{2\beta_x} \qquad (14.19)$$

The sum of all moments along the edges of the element has to fulfil the equilibrium equation for the element. When we know this sum and the moments taken by the middle strip we can easily calculate the moments in the column strip.

In determining the average total moments for the element we may take into account the higher stiffness of the solid part. This can be done by choosing a somewhat higher numerical value of the support moment. A suitable increase is in the order 10-15 % compared to the values for a normal solid slab.

The solid part has a higher dead load than the joisted part. As this load is acting close to the support it mainly influences the support moment. It can therefore just be added to the support moment within the column strip.

Where two corner-supported elements meet at a line of zero shear force in a regular system, the lateral moment distributions can be made to agree. In other cases there may be different theoretical lateral distributions. A typical case is where a corner-supported element meets a one-way element, which is supported on its whole width. The theoretical distribution of span moments in the corner-supported element is rather uneven, with higher moments in the column strip, whereas the distribution in the one-way element is uniform.

This case can be analysed on the assumption that the element with a support on its whole width is not a true one-way element, but that the middle and column strips continue into this part. Half the load on the middle strip is transferred to the column strip, just like in the corner-supported element. The reinforcement in the joists at right angles is the same as in the parallel joists in the corner-supported element. It can be shown that this analysis leads to an acceptable distribution of span moments, even if it is not always rigorously correct. A rigorous solution can always be achieved by assuming that a smaller part of the load than one-half is carried in the direction of the middle strip. This assumption leads to an increase in moments in the joists at right angles, which hardly seems necessary in practice.

As the joists are supposed to take only bending moments, the lengths of reinforcing bars can be calculated by ordinary methods. In the solid part, where torsional moments are also acting, it is recommended that all the support reinforcement is carried to the boundary of this part. No bottom reinforcement is theoretically needed within the solid part, unless there is a positive moment at the boundary. In this case the corresponding reinforcement is carried to the support line. Fig. 6.3.7 in *Strip Method of Design* shows the theoretical design moment distribution for the case $\beta_x = \beta_y = 0.5$.

As will be demonstrated in the example below, this analysis leads to a rather uneven lateral moment distribution, with small moments in the middle strips. The resulting design is

279

safe with respect to ultimate load. There may be a risk of unacceptable cracks in the middle strips, particularly on the top surface. The moment distribution is a consequence of the assumption that the joists can take only bending moments and no torsional moments. A lateral redistribution of design moments as for solid slabs cannot be made without the introduction and acceptance of torsional moments in the joists. Maybe some such redistribution can be accepted. Here, however, only the theoretical solution without torsional moments in the joists will be used.

Example 14.3

The joist floor in Fig. 14.3.3 has joists at 0.6 m centres. It has solid parts corresponding to 6 filled modules in each direction. The total widths of the column strips are thus $7 \times 0.6 = 4.2$ m in each direction. The supporting columns have square sections 0.5 m$\times$0.5 m. The outer boundaries are simply supported. The total load on the major parts of the floor is 11 kN/m^2. Within the solid parts the excess load is 5 kN/m^2.

We start by determining the average support moments in order to calculate the c-values, which we need for the detailed analysis. As stated above, the solid parts are stiffer than the joisted parts, and this is here taken into account as a factor 1.15 for the support moments. The excess load on the solid part is not taken into account at this stage.

$$m_{xs} = -1.15 \times 11(8.5^2/12 + 6.95^2/8)/2 = -76.27 \tag{14.20}$$

$$m_{ys} = -1.15 \times 11 \times 7.55^2/8 = -90.14 \tag{14.21}$$

The c-values are calculated with Eq. (2.34) and the average span moments with Eq. (2.35)

$$c_x = \frac{6.95}{2} + \frac{76.27}{11 \times 6.95} = 4.47 \tag{14.22}$$

$$c_y = \frac{7.55}{2} + \frac{90.14}{11 \times 7.55} = 4.86 \tag{14.23}$$

$$m_{xf1} = 11 \times 4.25^2/2 - 76.27 = 23.07 \tag{14.24}$$

$$m_{xf2} = 11 \times 4.47^2/2 - 76.27 = 33.63 \tag{14.25}$$

$$m_{yf} = 11 \times 4.86^2/2 - 90.14 = 39.77 \tag{14.26}$$

Noting that the width of the column strip belonging to each element is the distance from the corner of the column to the dividing line between column and middle strip, we find that

$$\beta_x c_x = \beta_y c_y = 3.5 \times 0.6 - 0.25 = 1.85 \tag{14.27}$$

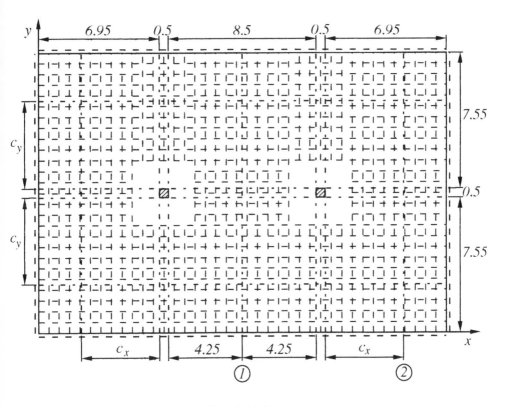

Fig. 14.3.3

$$\beta_{x1} = 1.85/4.25 = 0.435 \qquad (14.28)$$

$$\beta_{x2} = 1.85/4.47 = 0.414 \qquad (14.29)$$

$$\beta_y = 1.85/4.86 = 0.381 \qquad (14.30)$$

For the support moment in the middle strip in the x-direction we find somewhat different values from the two spans from Eq. (14.18). We use the average of these values

$$m_{xsm} = -11[(1 - 0.435)4.25^2 + (1 - 0.414)4.47^2]/16 = -15.07 \qquad (14.31)$$

Applying Eq. (14.17) we find the corresponding span moments in the middle strip

$$m_{xfm1} = 11(1 - 0.435)4.25^2/4 - 15.07 = 13.00 \tag{14.32}$$

$$m_{xfm2} = 11(1 - 0.414)4.47^2/4 - 15.07 = 17.13 \tag{14.33}$$

We can now calculate the average moments in the column strip in the x-direction by subtracting the part taken by the moments in the middle strips from the total average and distribute the result on the width of the column strip. For the support moment we now add the moment caused by the excess load on the solid part. Noting that this excess load is only acting on a width of 1.55 m of the total width 1.85 m of the column strip we find that the average excess moment is

$$\Delta m_{sc} = -\frac{1.55}{1.85} \times \frac{5 \times 1.55^2}{2} = -5.03 \tag{14.34}$$

$$m_{xsc} = [-76.27 + (1 - 0.381)15.07]/0.381 - 5.03 = -180.73 \tag{14.35}$$

$$m_{xfc1} = [23.07 - (1 - 0.381)13.00]/0.381 = 39.43 \tag{14.36}$$

$$m_{xfc2} = [33.63 - (1 - 0.381)17.13]/0.381 = 60.44 \tag{14.37}$$

It is also of interest to calculate the moment at the boundary between the solid part and the joists. For this purpose we calculate the load on the joist part of the column strip from Eq. (14.18)

$$q_{xc} = \frac{11 \times 1.381}{2 \times 0.381} = 19.94 \tag{14.38}$$

The moments at the boundaries closest to sections *1* and *2* respectively are

$$m_1 = 39.43 - 19.94 \times 2.70^2/2 = -33.25 \tag{14.39}$$

$$m_2 = 60.44 - 19.94 \times 2.92^2/2 = -24.57 \tag{14.40}$$

As these moments are negative some top reinforcement is required in the joists in the vicinity of the solid part. No bottom reinforcement is theoretically needed in the solid part. For the y-direction we get the moments in the middle strips from Eq. (14.18)

$$-m_{ysm} = m_{yfm} = 11(1 - 0.381)4.86^2/8 = 20.10 \tag{14.41}$$

The moments in and the load on the column strip have somewhat different values on both sides of the column, as we have different β_x-values. We may directly calculate the average value

$$m_{ysc} = -[(90.14 - (1-0.435)20.10)/0.435 + (90.14 - (1-0.414)20.10)/0.414]/2 - 5.03$$
$$= -190.22 \qquad (14.42)$$

$$m_{yfc} = [(39.77 - (1-0.435)20.10)/0.435) + (39.77 - (1-0.414)20.10)/0.414]/2 = 66.47 \quad (14.43)$$

$$q_{yc} = \frac{11}{4}\left(\frac{1.435}{0.435} + \frac{1.414}{0.414}\right) = 18.46 \qquad (14.44)$$

The moment at the boundary of the solid part is

$$m = 66.47 - 18.46 \times 3.31^2/2 = -34.65 \qquad (14.45)$$

In addition to the corner-supported elements, there is a thin one-way element at each side of the column. The moments in this element can be calculated separately and the corresponding reinforcement added. It is simpler, and slightly on the safe side, to assume that the moments in this element are the same as in the column strips.

The joists must be reinforced for moments in kNm corresponding to the relevant m-values in kNm/m multiplied by 0.6 m. The solid parts should be reinforced for total moments in kNm corresponding to the above support moments multiplied by 4.2 m, which is the total width of the support strip, including the one-way element. This reinforcement may be uniformly distributed over the width of the solid part, which is 3.6 m plus the width of one joist, or with some concentration towards the support.

CHAPTER 15

Prestressed slabs

15.1 General

The strip method is based on the theory of plasticity, and the main relevant plastic property of reinforced concrete slabs is the yielding of reinforcement at ultimate load. Prestressed slabs are mainly assumed to function elastically and their plastic properties are limited. The strip method as applied to slabs with ordinary reinforcement cannot be generally accepted as a design method for prestressed slabs. There exist, however, situations where the strip method may be useful.

One possible way of applying the strip method is to use the most basic principle of the simple strip method and divide the slab into a number of narrow strips, each strip containing one tendon, which carries the load on the strip. In this case the plastic properties of the slab are unimportant.

It is also possible to use a mixed design, e.g. with prestressing tendons as a support band along a free edge, whereas the rest of the slab has normal reinforcement. In this way the deflection can be limited.

When the strip method is applied in connection with prestressing it seems most natural to use the principle of load balancing, in which the load on a strip is balanced by the force caused by the change in direction of a curved tendon. Only this approach will be discussed below, and the reader is expected to be familiar with it. However, the analysis is often made in terms of bending moments corresponding to these forces. The idea of load balancing is of importance for the arrangement and shapes of the tendons.

The examples of applications below are also simplified by assuming only a constant load and disregarding prestress losses due to creep, shrinkage and relaxation, and other effects which are normally taken into account in the design of prestressed structures. The intention of this chapter is only to show some possible ways of making use of the strip method for the design of prestressed slabs, not to give complete guidance for such a design.

15.2 The simple strip method for tendons

The direct application of the simple strip method for a slab with prestressing tendons is similar to its application to joist floors. In a joist floor each joist carries the load from a certain part of the floor. The joists are at equal spacings and the amount of reinforcement varies depending on the load. In a prestressed slab all tendons are often equal and in order to utilise them the widths of the strips have to vary. In simple cases the widths of the strips are chosen to give them equal moments.

Example 15.1

The slab in Fig. 15.2.1 is simply supported. It has a load of 11 kN/m^2 and is to be prestressed with tendons, able to provide each for a moment of 15 kNm.

According to the principles of the simple strip method the slab is divided by means of lines of zero shear force into areas which carry the load in different directions.

The total moment to be taken by the tendons in the x-direction is, according to Eq. (2.4),

$$M_x = 6.0 \times 11 \times c_x^2/6 = 11c_x^2 \tag{15.1}$$

The number of tendons to take this moment is denoted n_x. We get

$$15n_x = 11c_x^2 \tag{15.2}$$

The total moment to be taken by the tendons in the y-direction is, according to Eq. (2.5),

$$M_y = 11 \times 3.0^2[8.0 + 2(8.0 - 2c_x)]/6 = 66(6.0 - c_x) \tag{15.3}$$

With the number of tendons to take this moment denoted n_y, we get

$$15n_y = 66(6.0 - c_x) \tag{15.4}$$

We can thus assume a value of n_x and calculate c_x from (15.2) and n_y from (15.4). If we assume $n_x = 5$ we get $c_x = 2.61$, $n_y = 14.9 \Rightarrow 15$. The theoretical total number of tendons is 19.9. It proves that we get approximately the same theoretical total number of tendons if we assume $n_x = 6$, 7 or 8. If we assume $n_x = 4$ we get a theoretical total number of 20.1.

In this case it seems suitable to choose $n_x = 5$, $n_y = 15$, $c_x = 2.61$.

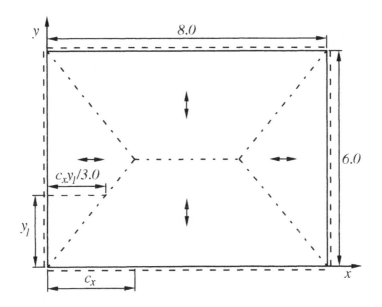

Fig. 15.2.1

Suitable positions of tendons are determined in the following way. The first tendon in the x-direction takes the load on a width y_1. The moment relation gives

$$y_1 \times 11 \times (2.61 \, y_1/3.0)^2/6 = 15 \qquad (15.5)$$

We find $y_1 = 2.21$. With the triangular load distribution within that part it is reasonable to place the first tendon at $y = 1.5$ or 1.6. A more exact determination of the position is not important for the behaviour of the slab.

We continue by determining a value y_2 for two tendons by replacing y_1 with y_2 and doubling the left side in the equation. This gives $y_2 = 2.79$. The second tendon thus takes the load between $y = 2.21$ and $y = 2.79$. It may be placed at $y = 2.5$ or 2.6. The third tendon is placed at $y = 3.0$ and the fourth and fifth symmetric to the second and first.

In the same way we find that the tendons in the y-direction may be placed at $x = 1.3, 2.1,$ $2.5, 2.8, 3.1, 3.4, 3.7, 4.0$ etc.

The shape of the tendons should be determined from the load that they are expected to carry. Thus, the tendons should be straight within the parts where the load is taken by the tendons in the direction at right angles. The first tendon in the x-direction should, for instance, be curved approximately in the first 1.3 m from the edge and then straight until the same distance from the opposite edge.

This design is the most economical with respect to the number of tendons. If, for some reason, it is thought better to have some other distribution of tendons, the load may be assumed to act in some other way, e.g. divided between the two directions within certain parts of the slab.

15.3 Prestressed support bands

In order to limit the deflection of a slab and prevent excessive cracking, it may be advantageous to use prestressing tendons in a support band whereas the rest of the slab has normal reinforcement. A typical case is where a large slab has one long free edge and the other edges simply supported. With only normal reinforcement the deflection may be unacceptably large. According to the theory of elasticity, much of the load is carried by torsional moments, which may cause large cracks in under service conditions, as reinforcement parallel to the edges is less effective for limiting skew cracks. With a prestressed support band along the free edge the slab may be made to act as if it is supported along also this edge and the load is carried mainly by bending moments, for which the reinforcement is effective for crack limitation.

Example 15.2

The slab in Fig. 15.3.1 carries a load of 11 kN/m². It has one long free edge and the other edges are simply supported. A support band with prestressing tendons is arranged along the free edge. The width of this support band, which depends on the type and number of tendons and on the concrete stresses, is assumed to be 0.6 m.

The slab is assumed to be supported along the centreline of the support band. This assumption is not the same as that normally used in this book. The reason for this difference is that the strip method is based on the theory of plasticity, which is not applicable for the prestressed support band. In order to be on the safe side in the case of a prestressed support band, this different approach is used. The difference is mainly one of principle and has a very small influence on the resulting design.

The slab is thus assumed to have a span of 3.7 m in the y-direction and the ordinary reinforcement is designed according to the methods in Chapter 3.

The support band has to carry the load from the slab plus the direct load outside the centre of the band. By means of Eqs (2.3) and (2.4) we find the moment in the support band

$$M = (3.7/2 + 0.3) \times 11 \times 3.5^2/2 - 1.85 \times 11 \times 1.5^2/6 = 137.2 \tag{15.6}$$

288

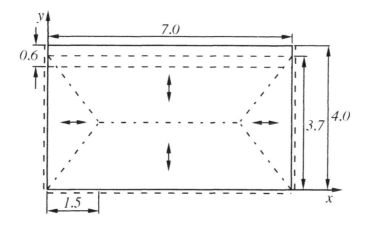

Fig. 15.3.1

The shape of tendons should be determined from the load distribution. As this is nearly uniform the tendons may be given a purely parabolic shape.

Even though the tendons balance the forces, this does not guarantee that the edge does not deflect. The uneven compressive stress distribution may cause curvature. This can be neutralized by suitable choices of anchorage, prestress force and shape of the tendons. The reader is referred to the literature on prestressed concrete.

15.4 Flat slabs

The analysis of prestressed flat slabs may, in principle, be made with the same approach as for joist floors. If the column strip can be made so narrow that it is not much wider than the supporting columns, such a design can be regarded as rigorously correct. With wider column strips the solution is less rigorous, as torsional moments appear within the area where the support strips cross. These torsional moments are necessary to bring the load to the support. The wider the column strips, the more important are the torsional moments.

The torsional moments correspond to principal moment directions which form an angle to the directions of the tendons. In the vicinity of the support the principal moments are mainly polar, with the tangential moments having the largest negative values, whereas the radial moments have smaller negative values or even positive ones. From equilibrium conditions it will be found that there is always a difference between radial and tangential moments

289

close to the support of a corner-supported element equal to $2R/\pi$, where R is the support reaction from the element. The slab has in principle to be designed for both moments.

The prestressed tendons cause a moment field in the slab with principal directions corresponding to the directions of the tendons. The resulting moment field acting on the concrete section is thus a combination of two fields having different principal directions. This resulting moment field has principal directions which differ from point to point, and at each point it has two principal moments which are different. The section should be checked with respect to both these moments at all points within the support area.

The stress situation is evidently extremely complex. A correct solution should be based on a moment field from the acting load calculated with the theory of elasticity, which unfortunately gives very uneven moment distributions, not directly applicable for design. As an approximation on the safe side regarding ultimate load, the design may be based on the moment field for a corner-supported element according to the strip method. For such an approach the following procedure may be used:

1. Determine β-values for the corner-supported elements from the chosen widths of the column strips. These widths depend on the acceptable concrete compressive stress.

2. Calculate the moments in the middle strips by the same method as for joist floors.

3. Calculate the loads on the column strips. The column strip in one direction is regarded as the support for both the middle and column strips in the direction at right angles.

4. Determine the moments for the column strips based on these loads and the assumption that the strips are supported at the column.

5. Starting from these curves, determine design moment fields by adding the negative and positive Δm-values according to Fig. 15.4.1, which is valid for the x-direction. This results in two curves within the support area. Both curves are to be taken into account in the design.

This design is safe with regard to the ultimate limit state. Maybe it is too conservative. It is probably also acceptable from the point of view of cracking, although it is not based on moments according to the theory of elasticity. The ratio between span and support moments should be based on the principles of the theory of elasticity.

Example 15.3

Fig. 15.4.2 shows a corner-supported element with $c_x = 5.0$ m, $c_y = 4.0$ m. The widths of the parts of the column strips within the element are 1.0 m in both directions. This corresponds to $\beta_x = 0.20$, $\beta_y = 0.25$. The load is 12 kN/m^2.

Applying Eq. (14.17) we find for the middle strips

$$m_{xfm} - m_{xsm} = 12(1 - 0.20)5.0^2/4 = 60.0 \tag{15.7}$$

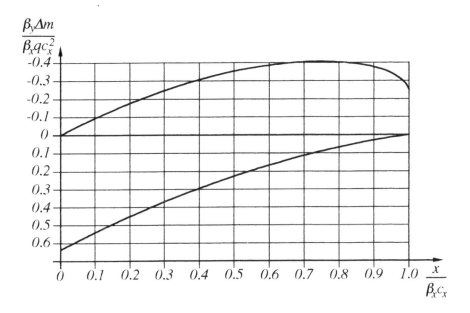

$$\frac{\beta_y \Delta m}{\beta_x q c_x^2}$$

Fig. 15.4.1

$$m_{yfm} - m_{ysm} = 12(1 - 0.25)4.0^2/4 = 36.0 \tag{15.8}$$

Applying Eq. (14.19) we get the loads on the parts of the column strips where they cross the middle strips

$$q_{xc} = 12 \times 1.25/0.50 = 30.0 \tag{15.9}$$

$$q_{yc} = 12 \times 1.20/0.40 = 36.0 \tag{15.10}$$

The moments in the column strips caused by these loads are

$$m_{xc} = m_{xfc} - 30.0(5.0 - x)^2/2 \tag{15.11}$$

$$m_{yc} = m_{yfc} - 36.0(4.0 - y)^2/2 \tag{15.12}$$

The loads on the column strips within the area where the column strips cross are

$$q_{xc} = 36.0 \times 3.0/1.0 + 12 = 120.0 \tag{15.13}$$

$$q_{yc} = 30.0 \times 4.0/1.0 + 12 = 132.0 \tag{15.14}$$

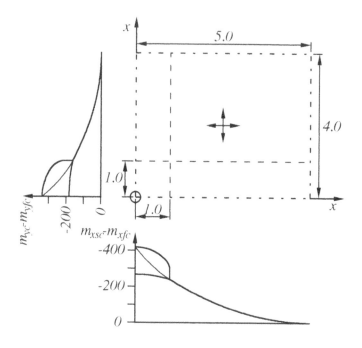

Fig. 15.4.2

The moments within this area are

$$m_{xc} = m_{xfc} - 30.0(5.0 - x)^2/2 - 90.0(1.0 - x)^2/2 \tag{15.15}$$

$$m_{yc} = m_{yfc} - 36.0(4.0 - y)^2/2 - 96.0(1.0 - y)^2/2 \tag{15.16}$$

The curves for the moments in the column strips according to these relations are shown in Fig. 15.4.2. The sums of moments in the column strips are

$$m_{xfc} - m_{xsc} = 420 \tag{15.17}$$

$$m_{yfc} - m_{ysc} = 336 \tag{15.18}$$

To the values according to the above relations are to be added the values from Fig. 15.4.1. The resulting curves are shown in Fig. 15.4.2. Both the upper and lower curves have to be taken into account in design. The concrete stresses resulting from these moments, combined with the influence of prestressing tendons, have to be checked.

References

Armer, G.S.T.: The strip method: a new approach to the design of slabs. Concrete, Sept. 1968, 358-363.

Crawford, R.E.: Limit design of reinforced concrete slabs. Journal of Engineering Mechanics Division, Proc. ASCE, Oct. 1964, 321-342.

Ferguson,M., Breen, J.E. and Jirsa, J.O.: Reinforced Concrete Fundamentals, 5th ed. 1988, John Wiley and Sons.

Hillerborg, A.: Equilibrium theory for reinforced concrete slabs (in Swedish). Betong 1956. 171-182.

Hillerborg, A.: Strip method for slabs on columns, L-shaped plates, etc. Translated from Swedish by *F. A. Blakey*, CSIRO, Division of Building Research, Melbourne 1964.

Hillerborg, A.: A plastic theory for the design of reinforced concrete slabs. IABSE Sixth congress, Stockholm 1960, Preliminary Publication, 177-186.

Hillerborg, A.: Strip Method of Design. A Viewpoint Publication, C&CA 1975. E & FN Spon.

MacGregor, J.G. Reinforced Concrete: Mechanics and Design, 2nd ed. 1992, Prentice Hall.

Nilson, A.H. and Winter, G.: Design of Concrete Structures, 11th ed. 1991, McGraw-Hill, Inc.

Park, R. and Gamble, W.L.: Reinforced Concrete Slabs, 1980, John Wiley.

Wilby, C.B.: Structural Concrete, 1983, Butterworth & Co. Revised as Concrete Materials and Structures, 1991, Cambridge University Press.

Wood, R.H. and Armer, G.S.T.: The theory of the strip method for the design of slabs. Institution of Civil Engineers, Proceedings, Oct. 1968, 285-311.

Index

Additional moment 7
Advanced strip method xiii, 10, 139, 144
Affinity law 34
Anchorage of reinforcing bars 45-47
  at a free edge 47
Approximations xiv, 2
Armer, G.S.T. xiv
Arrow, double-headed 11, 13, 15
Arrows, double-headed,crossing 29
Average moments 30
  corner-supported element 29-34
  one-way element 15-27

Band, reinforcement 40
Blakey, F.A. xiii
Breen, J.E. 9

Cantilever 24
Cantilevering flat slab 155-157, 196-211
Cantilevering slab 227
Capital 140
Circular opening 242
Circular slab 127-129
Column
  at a slab corner 154
  at obtuse corner 182

exterior 143, 151-155
  interior 143
Column capital 140
Column profile, theoretical 174
Column strip 31, 143
Column strip width 31, 139, 143, 280
Column support reaction 146, 148, 182
Combination of loads 60
Compatibility 9
Computer programs 5
Concentrated loads 35-37, 63-70, 82, 97, 112, 167-171
Concrete walls 269
Construction simplicity 27
Continuous slabs 257-270
Continuous strip 39
Conversion factors xi
Corner column 154
Corner element 39, 145, 215
Corner levers xv
Corner reinforcement 6, 73, 100, 112
Corner, reentrant 215-218
Corner-supported elements 29-34, 140, 213
  anchorage of reinforcing bars 47
  average moments 29-34
  design bending moment 47
  design moments 29-34
  joist floor 277-283
  length of reinforcing bars 47
  moment distribution 29-34
  non-orthogonal reinforcement 34
  non-rectangular 33

non-uniform loads 34
opening 255–256
point of support 140
rectangular
    concentrated loads 37
    uniform loads 30–32
support reaction 49
torsional moments 47
triangular 33
Corrosion of reinforcement 5
Crack control 6, 43
Crack limitation 5
Crack widths 44
Cracks 2, 5–7
Crawford, R.E. xiv
Creep 286
Crossing double-headed arrows 29
Curved edge 130, 187, 192
c-values 38, 40, 44, 45, 147

Deflections 2, 7
Deformations 7
Design moments 27–34
    corner-supported element 29–34
    one-way element 27–28
Design simplicity 27
Distribution reinforcement 35
    width 36
Dividing line 11
Double-headed arrow 11, 13, 15
Double-headed arrows, crossing 29
Drop panel 139

Earth pressure 77, 109
Economy 10, 11
Edge
    curved 130, 187, 192
Edges
    non-orthogonal 113–138
Elasticity, theory of 1, 4–5
Elements 12
    corner-supported 29–34
    one-way 13, 15
        average moments 15–27
    triangular at corner 39
    triangular one-way 15
Elliptical slab 130–134
Equilibrium equation 3, 9
Equilibrium theory xiii
Exterior column 143, 151–155

Fatigue 8
Ferguson, M. 9
Finite element analysis 4
Flat plates *see* flat slabs
Flat slabs
    average design moments 141–143
    cantilevering 155–157, 196–211
    column support reaction 146, 148
    crack limitation 143, 144
    design procedure 144–146
    exterior column 151–155
    individual strip 143
    irregular *see* Irregular flat slabs
    lateral moment distribution 144

lateral reinforcement
distribution 143
moment transfer to column 143,
144, 153
oblong panels 158–160
opening 255–256
prestressed 289–292
regular *see* Regular flat slabs
span 141
span length 140
strip width 141
strips 141
torsional reinforcement at exterior column 144
triangular corner element 145
width of column strip 143
Free edge, anchorage of reinforcing
bars 47

Gamble, W.L. 9

High strength concrete 2
Holes 231

Individual strip 143
Interior column 143
Irregular flat slabs 173–211
cantilevering 196–211
column support reaction 182
design procedure 174–176
edge curved and column
supported 192–196
edge curved and fully
supported 187–192

edges straight and fully
supported 176–182
edges straight and partly column supported 182–187
free edges 176
lines of zero shear force 174
span line 174
support line 175
theoretical column profile 174
Irregular one-way element 18
Irregular slab 10, 130–138

Jirsa, J.O. 9
Joist floors 271–283
column-supported 277–283
corner-supported element 277–283
lateral moment distribution 271
non-corner-supported
all edges supported 272–275
one free edge 275–276
torsional moments 271, 277

Large opening in supporting
wall 218–230
Lateral moment distribution 5
corner-supported element 29–34
flat slab 144
one-way element 12, 15, 27
Lateral reinforcement distribution
one-way element 28
Length of reinforcing bars 45–47
Lift at corner 7

Line
   dividing 11
   of zero moment 24, 100
   of zero shear force 11, 113, 114
Line loads 35
Liquid pressure 77
Live loads 7
Loadbearing direction 11
Loads
   close to a free edge 82–84
   combination 60
   concentrated 35–37, 63–70, 82, 97,
         112, 167–171
   concentrated and distributed
         together 68
   distribution between
         directions 45
   line 35
   linear variation 20–22
   on support band or beam 49
   point 35–37, 63–70
   shear force 24
   triangular 20–22, 57–62, 77, 109
   uniform 16–18
Lower bound theorem 2, 9
L-shaped slab 215–218

MacGregor, J.G. 9
Middle strip 31, 143
Minimum reinforcement 8
Moment
   additional 7
   average 30
   design 27–34

Moment distribution 3
   between different directions 45
   lateral 5
      corner-supported
            element 29–34
      flat slab 144
      one-way element 12, 15, 27
   strip method 3
   theoretical 16
   yield line theory 3
Moment ratios 6, 32, 44–45
Moment transfer to column 143, 144

Nilson, A.H. 9
Non-orthogonal edges 113–138
Non-orthogonal reinforcement 34
Non-rectangular slab 113–138
   with opening 242–245

Oblong panels 158–160
Oblong slab 8
One single interior column 146–148
One-way elements 13, 15
   anchorage of reinforcing
         bars 45
   average moments 15–27
   concentrated loads 35
   design moments 27–28
   irregular 18
   length of reinforcing bars 45
   rectangular 20, 22
   trapezoidal 16, 20, 22
   triangular 16, 20, 22
One-way strip 10

Openings 231–256
    at a free edge 247–250
    circular 242
    corner-supported element 255–256
    flat slab 255–256
    small 231
    triangular 238

Park, R. 9
Plastic behaviour 2
Plastic properties 2
Plasticity, theory of xiii, 2, 3
    exact solution 3
    lower bound theorem 2
    upper bound theorem 3
Point loads 35–37, 63–70
Prager, W. xiii
Prestress losses 286N
Prestressed flat slab 289–292
Prestressed slab 285–292
Prestressed support band 288
Principal moments 6
Punching xv, 146
Punching design 1, 148
Punching failure 1

Ratios between moments 6, 32, 44–45
Rectangular one-way element 20, 22
Rectangular slabs
    all sides supported 51–70
        concentrated loads 63–70
        triangular loads 57–62
        uniform loads 51–56
        with opening 234–242

one free edge 71–86
    concentrated loads 82–86
    opening at the free edge 247–250
    opening not close to free edge 245–247
    triangular loads 77–81
    uniform loads 74–77
two adjacent free edges
    non-uniform loads 96–97
    uniform loads 87–96
    with opening 250–255
two free edges 87–97
two opposite free edges
    concentrated loads 87
    distributed loads 87
    with opening 250
Reentrant corner 215–218
Regular flat slabs
    concentrated loads 167–171
    different loads on panels 164–167
    non-uniform loads 161–171
    uniform loads 139–160
    uniform loads in one direction 161–164
Reinforcement
    at band support 42
    band 40
        width 42
    concentration along free edge 71, 233
    corner 6, 73, 100, 112
    corrosion 5
    curtailment 4, 45–47
    direction 15

directions, skew angle
    between 24
distribution 35
    strip method 3
    yield line theory 3
economy 3, 10, 11, 27, 44
lateral distribution
    flat slab 143
    one-way element 28
minimum 8
non-orthogonal 34
ratio 2
support 24, 25
torsional at exterior column 144
yielding 6
zone of zero reinforcement
    along support 69
Reinforcing bars
    additional length 46
    anchorage 45–47
    anchorage at a free edge 47
    length 45–47
    length, example 46
    minimum length from line of
      zero shear force 197
    skew angle between 24
Relaxation 286
Ribbed floors *see* Joist floors

Safety 2, 3
    strip method 2, 3
    yield line theory 3
Service conditions 2, 5, 27
Serviceability 5
Shake-down 7

Shear design 1
Shear failure 1
Shear force 24, 25, 38
    in support band or beam 49
    line of zero 11
Shear, design for xv
Shrinkage 286
Simple strip method xiii, 10–28
    for tendons 286-288
SI-units xi
Skew angle between reinforcing
    bars 24
Slabs
    cantilevering 227
    circular 127–129
    element 12
    elliptical 130–134
    holes 231
    irregular 130–138
    irregular shape 10
    L-shaped 215–218
    non-orthogonal edges *see* Slabs
      with non-orthogonal ed-
      ges
    non-rectangular 113–138
      with opening 242–245
    oblong 8
    openings 231–256
    prestressed 285–292
    rectangular *see* Rectangular
      slabs
    triangular 99–112, 242
      with opening 242–245
Slab systems 257–270

Slabs with non-orthogonal
edges 113–138
four straight edges 114–127
all edges supported 114–116
one free edge 116–120
two adjacent free edges 122–
127
two opposite free edges 120–
122
Span in a column-supported
slab 140, 141
Span line 174
Span moment 40
relation to support moment 44
Span strip 24
Steel ratio 2
Strip 10–14, 38–40
column 31
continuous 39
in flat slab 141
individual 143
loaded only at one end 14
middle 31
one-way 10
span moment 40
support moment 40
Strip method xiii, 1–7
advanced xiii, 10, 139, 144
moment distribution 3
reinforcement distribution 3
safety 2, 3
simple xiii, 10–28
Strip Method of Design xiv, 1, 9, 30,
40, 71, 73, 101, 127, 213, 279
Support bands 10, 40–43, 71, 123

at obtuse corner 182
bending moment 49
between interior columns 173
design 76, 81
load 49
prestressed 288
shear force 49
width 40, 42
Support line 175
Support moment 40
continuous slab 257–259
continuous strip 44
fixed support 44
relation to span moment 44
Support reaction 48
column 146, 148, 182
corner-supported element 49
Support reinforcement 24, 25
Support strip 24
Supporting wall with a large
opening 218–230
Systems of slabs 257–270

Tank 78, 79, 266–269
Theoretical column profile 174
Theoretical moment distribution 16
Theory of elasticity 1, 4–5, 44, 290
Theory of plasticity xiii, 2, 3
exact solution 3
lower bound theorem 2
upper bound theorem 3
Torsional moments 4, 6, 10, 29, 47, 73,
87, 231
Trapezoidal one-way element 16, 20,
22

Triangular corner element 39, 145, 215

Triangular corner-supported element 33

Triangular loads 20–22, 57–62, 77, 109

Triangular one-way element 15, 16, 20, 22

Triangular opening 238

Triangular slabs 99–112, 242
   all sides supported
      uniform loads 101–104, 108–109
   concentrated loads 112
   corner reinforcement 100
   one free edge
      triangular loads 109–112
      uniform loads 104–108
   reinforcement directions 99
   reinforcement distribution 100
   with opening 242–245

Ultimate limit state 2, 3

Uniform loads 16–18

Units xi

Upper bound theorem 3

US-units xi

Wall 60
   triangular 109

Wall, supporting, with a large opening 218–230

Walls 269

Water pressure 57, 60, 77, 78, 79, 266

Water tank 78, 79, 266–269

Watertightness 269

Width of column strip 31, 139, 143

Width of support band 40, 42

Wilby, C.B. 9

Winter, G. 9

Wood, R.H. xiv

Yield line pattern 114

Yield line theory xiii, 1, 3–4
   moment distribution 3
   reinforcement distribution 3
   safety 3

yielding 6

Zero moment, line of 24, 100

Zero shear force, line of 11, 113, 114